The Magic of Amber

The Magic of
AMBER

Rosa Hunger

 Press Ltd
London EC1V 7QA

Published 1977

© NAG Press Ltd, London, England (in conjunction with Northwood Publications Ltd), and Rosa Hunger, 1977

Colour photographs and black and white illustrations not otherwise acknowledged by Studio Lorenzini

ISBN 7198 0061 7

Typeset in Great Britain by Bacchus Press Ltd, London, and printed and bound by T. & A. Constable Ltd, Edinburgh. Colour printing by The Grange Press Ltd, Southwick, Sussex.

Contents

To my sister, Amber Goldman

Acknowledgements
The author and publishers would like to thank the following
for permission to use illustrations:
    Christie, Manson & Woods: 81, 82, 84, 85, 99, 108;
Devises Museum: 23; Mary Evans Picture Library: 43; Royal
Pavilion, Art Gallery and Museums, Brighton: 21; Sotheby
& Co: 31, 46, 47, 50, 55, 64, 107; Victoria and Albert
Museum: 6, 34, 49, 53, 59, 60, 63, 93, 101, 119, 122.

List of Colour Illustrations

PREFACE

I should like to begin by giving my main reason for writing this book, which is simply that I love amber. It has always held a great fascination for me, and my knowledge has been acquired by virtually a lifetime of contact with it, as I joined my father in the family business at the tender age of fourteen.

The other very important reason is that so few books in English have been written about amber that it is pointless to search through the shelves of any bookshop in the country to find one. Those that did exist are either out of print or unobtainable. As a member of a business, Sac Frères of Old Bond Street, which specialises in amber, I have become increasingly aware that a book giving some basic facts about the substance, might be appreciated both by those who are already familiar with it and by others who would like to learn something of its origins and characteristics.

It has been of great assistance to me during my research into the history of amber to be able to consult two interesting and informative books by German authors whose work on amber has been of vital importance. I would like to express my gratitude to my friend Inge Goodwin for her invaluable help in translating this material from the German for me, and my appreciation of the many hours we spent together in enjoyable discussion of various other aspects of the book.

I would also like to thank the Victoria and Albert Museum and the British Museum for their unfailing courtesy and co-operation in giving me access to the priceless antique pieces of amber in their possession.

It is my sincere hope that the readers of this book will derive as much pleasure from perusing its pages as I have had in writing them.

London, 1977   *Rosa Hunger*

CHAPTER ONE

AMBER — ITS MYSTERY AND APPEAL

A mber is unlike any other semi-precious material. It is a warm tactile substance that possesses an aura of its own. There is a subtle mystery surrounding it, due to its almost timeless formation and existence, an existence of so many worlds buried beneath the earth and the sea, that it is beyond the imagination to visualise them. Two of the chief characteristics of amber are that it is never really cold to the touch, and that it is comparatively very light in weight. It certainly does, in its own particular way, give great serenity to the mind by its soothing contact with the fingers. These characteristics of warmth and lightness, combined with its ancient history, are an integral part of its beauty.

Amber varies in colour, showing every graduation from white to dark red and even black. Through the ages the various different colours have been valued according to changing fashions and the superstitions of many peoples as to their inherent protective or curative properties. According to Pliny, in ancient Rome the white or wax-coloured amber was regarded as worthless and only used for fumigation, while the transparent reddish amber was much prized. The clear golden yellow was the most favoured of all. During the Middle Ages, under the rule of the Teutonic Knights, on the other hand, the white amber was in the greatest demand, for the making of rosaries. Since the Renaissance, mosaics of all the different coloured ambers have often been used for a most decorative effect. The Baltic amber is mainly yellow, but there are so many shades of this that no two pieces seem alike. It can be completely clear or completely opaque, or an intriguing mixture of both.

Apart from the variations in colour and translucence, all types of amber contain odd marks and imperfections. These are usually small inclusions of biological material: fragments of leaves, tiny specimens of tree bark, and all kinds of insects (though these are rarer) — anything, in fact, which lay in the path of the resinous fluid as it flowed from the conifers.

Amber was obtained by fishing, dredging or mining. Fishing is probably the oldest method, apart from simply gathering what was washed up on the seashore. In Chapter Five I have described the different ways of collecting the amber, the tools and nets that were used, the special clothing that was necessary, and the risk to life and limb incurred during the whole process.

The working of amber makes use of technological methods derived from or closely connected with techniques for working glass, ivory and gemstones. There is a strong tradition, particularly during the Renaissance and Baroque period, of using ivory and amber together. A very early

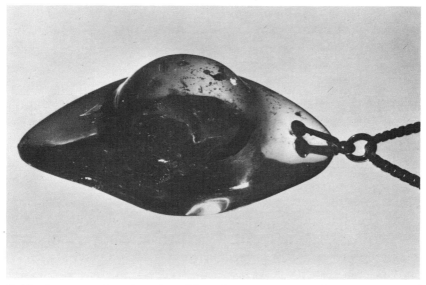

Golden-brown, unshaped pendant, 53mm long.

example of this was found near Kiev in Russia, a small lion's head of ivory with inset amber eyes which has been dated to about 800 B.C. Simple amulets of human or animal shapes were produced by Stone Age craftsmen, as were numerous amber beads, doubtless used for trading as well as for the adornment of the original Baltic inhabitants.

From the time of Hippocrates (around 400 B.C.), amber was also used as a medicine. Pliny's *Historia Naturalis* gives many weird and wonderful theories accepted in his time about this. Amber, he wrote, was not without its uses as a medicine, "though it is not only for this reason that women in particular are so attached to it"! Roman women believed that an amulet of amber worn around a baby's neck would protect it from all evils. According to the Greek writer Callistratus, it was good for people of any age, especially the clear golden stone which, worn round the neck, was a cure for fevers and other diseases. When ground into powder and

3

mixed with honey and oil of roses, it was said to be helpful for maladies of the ears or failing sight, and a remedy for diseases of the stomach. The oil obtained from it has been prescribed, to be taken internally, against asthma and whooping cough and also as a lubricant for the chest. In the Middle Ages it was thought to prevent all kinds of illnesses — the plague, heart disease and vertigo. Like many rare and expensive materials through the ages, it was said to cure impotence. Blood-stilling properties were attributed to it, which explains why a ritual Jewish circumcision knife of the eighteenth century had an amber handle.

In the sixteenth century amber was widely used as a medicine. It was ground into powder and swallowed in small doses, and was generally believed to be beneficial in the treatment of asthma, hay fever, goitre and other diseases. It is of course possible that the supposed effects were purely psychological. I have, however, known a great many people who believed strongly in the healing powers of amber, and many others who regard it as a talisman. There could be some foundation for this, in that the electrical properties of the substance might help to keep cold and chills at bay, as it takes and retains the warmth of the body for quite some time after it has been worn.

While these theories of amber as a medicine do not carry much weight in the modern world, the belief in its powers as a good luck charm still holds. Amber beads are extensively used in Greek and Arab lands as prayer necklaces and worry beads. Their lightness enables them to be carried easily in the pocket, ready to be taken out and passed through the fingers whenever desired.

Of course amber has always been of great importance to the smoker, in cigarette holders, mouthpieces and pipe stems. It is extremely pleasant used in this way, and exerts a

86 bead necklace of variegated amber with a fine carving as pendant.

Gamesboard decorated with bone plaques and verre eglomisée, 17th century.

1. Master snuff bottle, with carved and painted spatula, Chinese, 77 mm high without lid.

2. *Overleaf:* Spider and web inclusion in clear Baltic amber, set as a pendant; greatest dimension 57 mm x 33 mm.

cooling influence on the smoke that passes through it. Some of the most beautiful pieces of opaque amber have been used as mouthpieces for Turkish hookahs. These include a vessel for water or rose water through which the smoke passes and is thereby cooled and purified. Many of these special pieces are so highly prized that they have been ornamented with precious stones set in or around them.

Obviously amber is a most versatile and absorbing subject that will always hold the interest, possessing elements of mystery and change that so many people find irresistibly fascinating. There is its romantic origin, the incredible length of its life, under the sea, battered by waves, thrown on to rocky shores. There is the fact that it is virtually impossible to find two pieces that are identical. Some distinguishing mark changes the character of each piece: a difference in the shape, a patch of cloud in an otherwise clear surface, all kinds of minute biological inclusions, a tiny fly or mosquito, even a flaw or a bubble.

There are the dozens of different colours, the white opaque amber, the red and the brown, the pale yellow, the deep gold, the honey coloured, the greenish tints and the very rare blue. The variety is endless, and all the colours can shine with a depth or brilliance that in some becomes quite dazzling. All amber in its rough state has an outer coating which must be removed, and the surface polished, before the interior may be seen.

There is the intriguing thought that it was the women of the Stone Age who first began to love and wear amber, while women today find it as fashionable as ever.

A great many people still know nothing about amber. Some may own and wear ornaments or jewellery of this lovely material and yet be wholly ignorant of its nature and its history. This is understandable, as so few books have been published on the subject. The purpose of the present book is to acquaint those who would like to know more with its characteristics, and to answer some of the questions that have been put to us during our long experience of trading and specializing in amber.

CHAPTER TWO

FORMATION AND EARLY HISTORY

Amber is a fossilised tree resin, the parent tree being a species of pine, *Pinus succinifera*. The trees grew, together with palm trees, camphor laurels, bay trees, oaks and yews in the early (Eocene) period of the Tertiary Formation, roughly fifty million years ago, on a then luxuriantly wooded part of the European mainland, which today is covered by the Baltic Sea. Distribution of the amber pine seems to have been largely restricted to this area. Generations of trees for many thousands of years exuded their resin into gigantic storage chambers in the ground until eventually, with the sinking of the land and the invasion of the sea, the tree trunks were swept away, and a new stratum was built up from the resin and other components of the soil.

From the scientific study of the fossil plants of that time, three distinct epochs of vegetation seem to emerge: the first produced ordinary black coal, the second produced brown coal, and the third of submarine forests. The source of the black coal was an exceedingly rich vegetation. This flourished originally on many islands which rose from the sea and then sank again, only to re-emerge later, each time becoming covered with an abundant plant life.

At that time a high and even temperature existed over the whole northern hemisphere, so that the same fossil plants are found in the coal seams of both North America and England. Throughout this long period, the general temperature of the earth's surface gradually decreased. More islands rose from the sea that covered what is now the plain west of the Ural Mountains. These islands, like the earlier ones, produced a complete flora but it was very different from that of the black coal period. The North European lowland, as it rose from the waves, became the

Fly in yellow amber set as ring, 34mm long.

home of the plants which eventually formed brown coal. This preserved a whole fossilised organic world much nearer our present one than that of the black coal period, though still distinct and primitive.

This epoch was also of great length, consisting of periods separated by the frequently recurring inundations from the north polar ice-cap. The formation of amber belongs to the intermediate periods of the brown coal age. The submarine forests which succeeded the brown coal age show a transition to a kind of vegetation closer to that of the present day. The investigation of organic remains preserved in amber has therefore proved of great importance in illustrating this particular part of the world's history.

Amber has been found in all the countries bordering the Baltic and the North Sea, including Great Britain, and as far away as Sicily, Central Europe, the Balkans, the Caribbean and even Burma; but the northern coast of Prussia has always been the greatest amber source of the world. On this Baltic shore, the peninsula of Samland going north-west from Königsberg (now Kaliningrad), was the centre of the famous amber-bearing dark sand and clay layer known as the "blue earth". Other layers containing amber were deposited above the blue earth. In later ages, gigantic glaciers descended from the North and swept away portions of the amber-bearing layers, scattering them over the North German lowlands. This explains why isolated finds of amber have been made in clay soils and sandpits all over Germany.

When the land upheavals eventually stopped and the countries surrounding the Baltic assumed their present form, deep down on the sea bottom the blue stratum was brought into contact with the water and in places even formed the surface of the sea bed. From then on it was the action of the water that altered the configuration of the soil. With the added effect of wind and vegetation and animal life, the blue earth was sucked out of the sea bed, or washed from the submerged coastline into the sea.

The scientific study of the zoology and botany of amber has only begun within the last fifty or sixty years. Very many students of natural history have been enabled to increase their store of knowledge greatly by studying the remains of flora and fauna entombed and preserved in amber.

These studies often prove of enormous interest with regard to the many small creatures existing in the amber forests at the time of the piece's formation. A number of these insects belong to quite different species and show different characteristics from those living today. On occasion an insect may combine the characteristics of several modern species. This is illustrated by the existence of one small creature which, from the structure of its antennae, feet and mouth parts, belonged to the order of lacewings (*neuroptera*), but shows forewings reminiscent of butterflies (*lepidoptera*).

Many insects have been discovered, studied and discussed among the amber fauna. A rough estimate would probably bring the number up to a thousand species. Among those found have been various larvae and caterpillars, bees, ants, flies, earwigs and butterflies, and among the non-insects have been found spiders, centipedes and small land snails. All these have been minutely examined.

During the nineteenth century, three German naturalists were foremost in the study of the remains found in amber. Around 1830 Dr. G.C. Berendt, in association with Professor H.R. Goppert and Professor A. Menge, collected and investigated over two thousand pieces of amber containing animal and vegetable specimens.

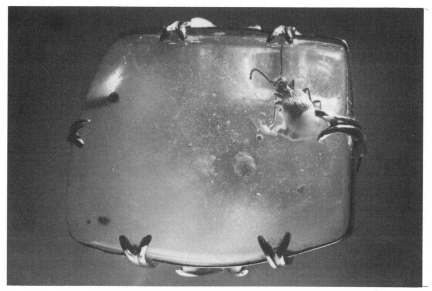

Beetle in pale yellow amber set as ring, 26mm × 20mm.

Numerous fragments and particles of leaves, flowers and other plant parts which had been broken or blown off and scattered about the forests by the wind, were caught in the liquid resin and held for posterity. From these clues all deductions regarding the flora of the amber forests had to be drawn. Among the trees and plants recognised and studied by the experts were beech, alder, birch, poplar, oak, willow, cypress, chestnut and camphor trees, also many kinds of moss and fungus.

An interesting and unusual suggestion with regard to the amber trees, *Pinus succinifera*, was put forward by Johann Christian Aycke in 1835. He was of the opinion that the trees must have been in a diseased condition to exude the resin in such excessive quantities. From his observations it would seem that, at times, whole portions of the solid wood became entirely transformed into amber fluid. The white opaque amber and the transparent amber often occur

13

together in one piece, either in layers with definite limits, or each merging into the other in the manner of a mixture of clear and muddy water. Aycke pointed out that both kinds must have poured from the tree at the same time, and from the same aperture. The resin was in various forms of consistency as is shown by the different shapes. The so-called "petrified pins" are especially interesting. They were formed from long, tough threads of the resin, and were kept in their original shape by a subsequent flow pouring over and enclosing them. These "pins" must have been produced by a viscous fluid thicker than that which entrapped the small insects such as midges, water moths, ants and termites. They were able to work their way into the central part of the fluid after being trapped by the resin.

Of the many insects found in their fossilised tombs, by no means all were complete. The degree of preservation of some, however, is quite miraculous; some even retain the colour on their wings and bodies. Others, not so fortunate, plainly show the result of their vain death struggles as the fluid enveloped them: the separate legs and wings lying near the small corpses are sufficient evidence of this.

Many fine specimens of amber containing insects can be seen in various museums all over the world. Private collectors also have their share. Here and there are pieces containing swarms of flies of many varieties. I have one small piece in my collection which shows fifteen of the little creatures. Clearly they were attracted by what seemed to be a sweet and sticky drink, but proved to be a deadly trap from which there was no escape.

More rarely found than the smaller insects are spiders in amber. I am fortunate in possessing one set right in the middle of a lovely clear gold piece of amber. It is very well preserved and the whole body can be seen in perfect clarity.

Many varieties of beetles have been discovered through the study of amber remains. One of the most interesting of these I have had set into a ring. This little fellow has been entombed in opaque amber and, when viewed under a magnifying glass, he is a magnificent sight. From his rather large head protrude two thick antennae. His back is covered in dark red stripes with what appears to be a small misty cloud surrounding the end of his body. I believe this to be a liquid ejaculated by the insect as a defensive measure when feeling the resin descend upon him.

It is a unique and utterly fascinating aspect of amber that it has provided the means by which scientists have been able to investigate the secrets of the past from the flora and fauna trapped in their fossilised tombs. There is no doubt that amber is the finest form of natural preservation the world has ever known.

I end this chapter with four charming and humorous lines from the "Epistle to Dr Arbuthnot" of Pope:

Pretty! in amber to observe the forms
Of hairs, or straws, or dirt, or grubs, or worms!
The things, we know, are neither rich nor rare,
But wonder how the devil they got there?

CHAPTER THREE

AMBER IN PREHISTORIC AND ANCIENT PERIODS

Amber is a material of the greatest antiquity. It was already much in demand long before the Christian era, and is still a great favourite among gemstones today. Among the countries of the world it has had many different names. The Greeks called it *Electron*, the Romans *Electrum*, *Succinum* or *Glesum;* in Arabic the word for it was *Ambare*, in Persian, *Karabe*, in French, *ambre*. The German name for it was originally *Bornstein*, this later became *Bernstein*. The literal meaning of this was "burning stone", as it would sometimes be used as a fumigant or form of incense and burnt for its aroma. Of course there are many other names for it in other languages, but I think we will leave it at that, and henceforth refer to it simply by its English name.

Thousands of years after the end of the last Ice Age, when the population of Europe was beginning to increase, primitive tribes came to live on the shores of the Baltic Sea. Beachcombing after a storm, they discovered that the sea had brought them a splendid gift. Violent storms beating against the coast swept the amber from the blue stratum and hurled it, amid piles of sea plants, on to the shore. Then from the wilder shores it would be swept away again, to collect and form large masses in the more sheltered places. The inhabitants of the coast, who added to a meagre living as farmers by fishing, found this glittering gift from the sea, and began to trade what more advanced peoples considered an object of luxury for tools and weapons. Thus by means of barter the foundations of the amber trade were laid, and the amber routes were developed.

These lay mainly along the rivers. One route followed the Memel to the Dnieper and went on to the Black Sea; another went up the Vistula over the Carpathian Massive to the Danube, and then either southwards through Macedonia to Greece, or on to the Black Sea again. A third route took the amber westwards to Jutland; the ancient Phoenician seafarers took that to be the country where amber originated. From Jutland some of the amber found its way south, probably via the Rhine, to the plains of North Italy. Here men from Northern Europe met the men of the Mediterranean, and the amber changed hands in individual barter.

It is possible that middlemen brought the amber from Jutland to North Italy.

Once again the place of interchange was mistaken for the place of origin. The ancient Greeks firmly believed the valley of the river Po, which they called the Eridanus, to be the primal source of their *electron,* and invented the beautiful legend of Phaeton to account for it (see Chapter 13, The Literature and Mythology of Amber). Mention of *electron* is to be found in Homer. The fact that it developed a charge, an attraction for small objects, on being rubbed was known to the Greek scientist Thales of Miletus, writing in the 6th century B.C. Thus the phenomenon of electricity became forever linked with the Greek name for amber.

The origin of amber presented a problem of the greatest interest to past ages. In ancient times, both in the narratives of the Phoenicians and in the Greek myths, many fantastic theories were offered about its formation. However, some at least of the classical writers realised that amber had begun life as a resin, though they often, mistakenly, selected the black poplar as the amber tree.

The best classical account of the subject is given by Pliny the Elder in *Historia Naturalis* (Natural History) a 37-volume encyclopedia. This famous Roman author, who wrote in the first century A.D., was possessed of an insatiable curiosity and a strong critical faculty. He was rightly sceptical of the many and varied theories about the origin of amber as expressed by the Greeks and understood that it was a petrified tree resin. Perhaps this was being too hard on what were essentially charming poetic legends rather than serious scientific pretensions.

This shows him to be well in advance of at least one eighteenth century German naturalist who believed it to be "condensed sea foam, mixed with a great deal of salt on the sea shore, which, dried by the air and the heat of the sun, attains its extraordinary hardness". Even the Greek legends came closer to the truth.

Pliny mentions two reasons for believing that amber was a product of pine trees: firstly, when rubbed, it gives out an odour of pine; and secondly, when set on fire, it burns after the fashion and with the scent of a resinous torch. It would not be wise for anyone other than an expert to tamper with treasured possessions, but it has been a fascinating experience when actually working on amber to catch occasionally a really deep odour of pine, and to realise that this material has retained the smell for millions of years.

Sophocles believed that amber was the tears of the Meleagrides, sisters of the hero Meleager, who according to legend were transformed into birds, as recalled in Thomas Moore's beautiful lines:

"Around thee shall glisten the loveliest amber
That ever the sorrowing sea-bird has wept."*

Nicias, quoted by Pliny in the *Historia Naturalis,* believed that the resin was of solar origin, directly generated by the rays of the sun.

Demonstratus maintained that it was a secretion of the lynx, and called it *Lyncurion.* There were many other suggestions. Some writers, including Aristotle, did believe that it was a tree resin, though they differed as to the kind of tree and the locale where it was found.

Amber beads and ornaments are found among the ancient remains of Egypt, Greece, Italy and many other lands. Magnificent examples can be seen in private collections and public museums all over the world. Considering that this material is of a brittle and somewhat fragile nature, the survival of these primeval works of art, buried deep through the centuries, seems astonishing.

With the exception of metals and ivory, no article of commerce can be traced as far back as amber. It was the search for tin and amber that first enticed the ancient seafarers of the Mediterranean into the wilder regions of the west and north of Europe. A thousand years before Christ, the Phoenicians were already trading in amber. It was

---

* *The lament of the Peri for Hinda* from *The Fireworshippers*

known to the Greeks, too, many centuries before the Christian era: this is clear from allusions to it made by their earliest writers.

It is interesting that some of the earliest Stone Age amber objects were found well to the south of its source of origin; in Lower Austria, France and the Pyrenees. This may mark the southward migration of Baltic peoples, carrying their treasures with them, rather than the existence of important trade routes. Large numbers of pendants, beads and buttons have been found at 2nd Neolithic sites in Estonia. These sites date from between 3700 and 3300 B.C. and may well be one of the earliest cultures to use amber.

Much of the amber trade in Iron Age times was in the hands of the Etruscans, and one of the great trade routes passed through northern Italy and the Po Valley. This explains the great abundance of amber in this region. In Pliny's time the women living to the north of the Po Valley commonly wore amber necklaces instead of metal collars, and Etruscan tombs are rich in amber ornaments.

To the north of the Alps, only the ancient Bronze and Iron Age sites at Hallstatt in Austria show evidence of a widespread use of amber. There it was evidently so plentiful that it was used extensively as an adornment for women. In the Hallstatt tombs, elaborately worked Etruscan ornaments were found in great quantity, and show a complete contrast to the cruder products of the barbarians themselves.

Numerous beads, discs, necklaces and other amber articles, found in graves of the Stone and Bronze periods, show some skill in perforating, cutting and polishing; but there were no native artistic productions, either along the North Sea or Baltic coasts or in Central Europe, to compare in splendour and precision of craftsmanship with some of

The cup from Brighton Museum, Second Bronze Age.

the Hallstatt finds. All this is strong proof of the early development of man's aesthetic sense, the source of all the arts and refinements of modern society.

One of the most important pieces disovered in Europe was the famous cup made entirely of amber which can be seen in the Brighton Museum. It was found in 1857 during an excavation underneath a grave when work was begun on the construction of the first Hove railway station. Workmen came upon a rough coffin which had been crudely shaped in wood, evidently using very primitive tools. In the middle of this coffin, amid the remains of decayed bones and certain ancient workmanlike objects, such as a dagger and an axe, believed to date from the Second Bronze Age, lay the cup. Carved in one piece out of a very large block of amber, it is 3½ inches/8.9 cm wide and 2½ inches/6.4 cm high, and quite plain except for two or three raised lines curving around the top as far as the handle. The colour is a reddish gold but the amber has lost most of its original lustre owing to its great age. Its size, splendid simplicity and workmanship which is amazingly advanced for the time, make it unique.

In England, most of the amber found in the Bronze Age sites comes from Wessex where the "Wessex Culture" was flourishing. This people had many contacts with both northern Europe and the Mediterranean. They were associated with the building of Stonehenge III. Their barrows, or burial mounds, contain much gold, ivory and amber. The amber is usually in the form of necklaces or set in a gold circle: and it has been suggested that the latter may have been solar amulets. A fine necklace of twenty-two large graduated beads was found in a barrow in the New Forest near Beaulieu. This was dated to between 1800 and 1500 B.C.; this is contemporary with the Wessex Culture.

A grave in the cemetery at Upton Lovell in Wiltshire produced an amber necklace of 1,000 beads. A woman's grave produced a 270 bead necklace; this was at Long Wittenham in Berkshire. A single gigantic bead was found in the Upchurch cemetary in Kent; it measures 2½ × 1½ inches (63 × 38mm).

A rare find of amber in Ireland came during excavations at Tara, the home of the ancient High Kings of Ireland. While investigating the "Mound of the Hostages", they came upon the grave of a 14-15 year-old boy who was wearing a necklace made up of jet, amber, bronze and faience beads. By his side they found a small bronze knife; obviously he was buried with his two favourite possessions.

The Romans may have known amber as early as the time of the Kings of Rome. Early in the nineteenth century on Monte Crescentio near Marino, a discovery was made of a number of terracotta urns of very rough workmanship,

22

3. A swarm of small flies forms the inclusion in this piece of amber, 27 mm long, mounted in a gold ring.

4. A Chinese snuff box with carved panels, the lid carved with a rampant dragon, 54 mm across.

surrounded by yellow volcanic ashes under a layer of volcanic rock some five feet thick. In each urn was the model of a strangely shaped hut, also made of terracotta, and contained in each model were the charred remains of human bones, accompanied by various vessels and other articles of amber and bronze.

In 1667 a famous Italian botanist, Paolo Boccone, described some ancient tombs found near Ancona. The coffins were of stone, and in one of them strings of amber beads were found in extraordinary profusion, some as large as hens' eggs.

An interesting find was made in Rome in 1890 during excavations of the foundations for the Place of Justice in the old gardens of Domitian. Two sarcophagi were discovered, one of which was the tomb of a young bride named Crepereia Tryphaena. She was represented asleep on her funeral couch, at the foot of which stood two mourners, man and woman. Inside the coffin were the remains of the corpse, the head still covered with beautiful long hair. Among the treasured objects put into the coffin with the dead girl were a delicately carved wooden doll, several rings, and an amber hairpin. A well-preserved crown of myrtle leaves was probably her nuptial wreath. This sarcophagus, appealing equally to the archaeologist and to human sympathy, seems to date from the 2nd century A.D., and was placed in the Museo Capitolino.

In every age artists have worked with this rare and valuable material whenever they could get hold of it. An object of veneration in prehistoric times, in classical times it was already used for sculpture and bas-reliefs, especially under the Roman emperors. The smaller pieces were classed as precious stones and made into necklaces or set in gold or silver.

Amber was very much the fashion in Ancient Rome. When the Emperor Nero, a man known for his luxurious tastes, likened the colour of his wife's hair to amber, this was sufficient to bring the gem into immediate favour with all the Roman ladies.

Pliny tells also of the Roman knight who was sent to Germany to procure amber for the gladiatorial games given by the Emperor Nero. This knight travelled to the markets and shores of the country and brought back such an immense quantity of amber that the nets intended to protect the public from the wild beasts, were studded with it. The arms and the biers and all the equipment for the Games were similarly adorned for the entertainment of the Emperor and the people of Rome.

After existing for over a thousand years from prehistoric times, the amber trade between Italy and the Baltic began to fade, and finally died out entirely. Many different reasons have been given for the decline, but the migration of Teutonic peoples after the formation of the Eastern Roman Empire is generally accepted as being the main cause.

Kohn, the German translator of Sadowski, suggested that it could have been the strong influence of fashion, always fickle, that dealt the most serious injury, if not the death blow, to the trade. With reference to the great quantities of amber circulating through the classical world in Pliny's time, Kohn asked, "What would become of the diamond trade if the diamond fields were to yield their riches in such quantities that all the maidens of all the villages were able to adorn themselves with necklaces of the brilliant stones?" However, in a half-civilised country, the jewellery of the women often represents the savings of an entire family, and consequently the existence of large stocks of amber articles belonging to the peasantry

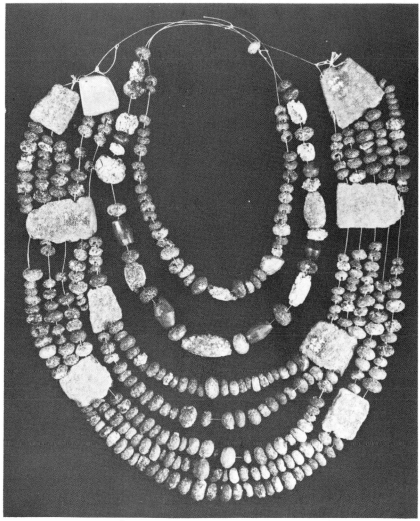

Necklace from Upton Lovell cemetery, Early Bronze Age.

Dagger pommel from Early Bronze Age barrow on Hamelton Down, Devon. Decorated in pointillé technique with gold pins, 58mm × 33mm.

Amber beads form part of this Anglo-Saxon necklace from Abingdon.

would not by itself prove that amber had lost its value.

According to Kohn, further damage was done to the amber trade by its transformation into direct trading, whereas in previous centuries it had taken the form of a system of barter. The bright Roman coins used as the original currency were at first gladly accepted by the northern barbarians, but later when the coins had been debased, the amber sellers were less ready to take them. Confidence is more easily destroyed than restored, and in the war-torn Europe of the Dark Ages much of early civilisations was lost or submerged for a time.

Amber ornaments found in graves in Great Britain are very similar to those found elsewhere in Europe: the shape of the beads in the necklaces is exactly the same as at Hallstatt nearly 1,000 years earlier. Such ornaments may well have been imported, rather than made in this country. Even on the amber coasts of the Baltic and North Seas, the rough material was probably exchanged for the finished article wrought by specialist craftsmen elsewhere along the amber routes.

Amber beads have been found in many Anglo-Saxon cemeteries throughout the country. The belief that amber possesses a power against witchcraft and evil spirits is the most credible explanation for the presence of one single bead in an Anglo-Saxon grave. Sometimes the piece of amber found was obviously badly worn or broken; clearly so long as it was amber, the condition was less important. One tomb near Sandwich contained a silver ring set with a lump of amber.

Anglo-Saxon cemeteries vary in the number of beads they produce. They may range from a single amber bead which forms the centre-piece of a necklace of glass beads, to a complete necklace.

T.C. Lethbridge, the eminent archaeologist, points out that amber is scarce in early 6th century graves, but becomes more common in those of the second half of that century. He draws the inference that the material could be of local, rather than Baltic, origin. He wondered whether possibly "the right to collect amber from the foreshore was once vested in the ruler of East Anglia and that amber was traded away for some of the many foreign goods imported into that area. The great amber forest seems to have its western boundary not far off the coast of England, and the East Anglia folk may therefore have had a monopoly in the trade."

Later graves in the Abingdon Cemetery contained more amber beads than the earlier ones. This cemetery lies on the Icknield Way, one of the main trading routes of England.

Many different objects of all kinds fashioned out of the precious material have been deposited in graves everywhere through the ages: jars, cups, medallions, rosaries, mirrors, knives and forks and other household articles. The lovely technique of amber-and-ivory inlay on handles can be seen as early as the Hallstatt remains. The custom of burying the treasured possessions with the departed, bears witness to how highly amber was prized. It is satisfying, too, that there can have been no other commodity that played a more active part in establishing European culture and skills with the rest of the ancient civilised world.

In conclusion it is interesting to note that all the fascinating things that have been found in tombs were in very varied states of preservation. Inevitably some have been affected by chemical processes in the soil over a period of centuries. Whatever source amber comes from, once it is cut and polished, it is subject to all the attacks of age and substances in the environment.

CHAPTER FOUR

AMBER IN THE MIDDLE AGES AND RENAISSANCE

During the medieval period individual ornaments in amber were rare. There was a very good reason for this: the supply of amber throughout the Middle Ages was only just sufficient to meet the Occidental demand for rosaries.

With the conquest of Prussia by the Teutonic Knights during the second half of the thirteenth century, production and trading in amber assumed a definite form. Like the Baltic Dukes of Pomerania before it, the Order of Teutonic Knights tried to enforce measures to secure exclusive exploitation of all finds of amber for itself. They claimed it as a Royal prerogative and assigned the rights of collection to individual nominees: thus in 1264 the agent for the Order was the Bishop of Samland; in 1312, the fishermen of Danzig; and in 1342, the Monastery of Oliva. The exclusive rights of sale developed more gradually in the course of the fourteenth century, and were finally confirmed in 1394 when, in the three towns that made up the city of Königsberg (Altstadt, Kneiphof, and Lobenicht), it was forbidden for any citizen to possess unworked amber.

The inhabitants of the shore, mainly fishermen, were under an obligation to collect the amber swept up on the shore, and to net it or dig it up from the shallow sea. All of it was then delivered to officials of the Teutonic Order, the Amber Lords, who paid for it in money and salt.

All branches of the amber trade came under the direction of the Grand Steward of the Order at Königsberg. He gathered in the raw material, paid the Amber Lords, paid commission to the Bishop of Samland and the Abbot of Oliva, then sold the disposable amber to Bruges and Lübeck where the paternoster-makers, united in powerful enclosed guilds, made it into rosaries.

Through this trading organisation, amber became the chief source of income for the Order in the Eastern region. It paid for the upkeep of the various Houses of the Order and subsidised its other widely diversified commercial enterprises. The Order was soon able to transform the natural wealth which providence had dealt out to it, into vital cultural assets as well, undisputed by any power or rival in the then known world.

Uninterrupted collection of all amber finds was essential to safeguard this most valuable source of capital. As they took delivery of the treasure, the Amber Lords had simultaneously to educate the shore dwellers in the discipline of the finding procedures. The shore-keeper would act ruthlessly against any attempt at theft: any unauthorised person caught collecting amber was strung up on the nearest tree. In one Prussian folk tale, the cruel shore-keeper takes a prominent role: on stormy nights when the sea throws up a particularly large amount, promising

Figure of an apostle, German, 125mm high, 16th century.

good pickings for the next day, the keeper's malevolent spirit was said to walk abroad, wailing, "Oh God, free amber, free amber!"

Apart from severely punishing dishonesty in delivering up all amber finds, the Order effectively countered any theft of amber by prohibiting any amber craftsman from settling in its territory. This made the working of amber illegal throughout the region, including Königsberg and Danzig, and ensured that all the raw material went to Bruges and Lübeck, the only towns where there were guilds of amber-turners. There were no amber workers' guilds in the Eastern region.

When a guild of amber-turners was formed in Danzig about 1480 (after the Teutonic Knights had fallen under the feudal rule of the King of Poland), the Order objected to this and sent a formal letter of complaint to the King. In time, however, it became convinced of the futility of protest, and eventually decided to inlcude the city of Danzig in its trading organisation, entering into an agreement to supply amber to the citizens of Danzig in 1483.

The strict ban on the working of amber did not mean that the Order completely rejected any artistic use of the material. As an exception to the rule the knights frequently ordered ornamental works, both for the Grand Master's own Court and for gifts to honoured outsiders. They even appear to have employed a special amber carver of their own. The name of one such sculptor was Master Johann, and he lived in Königsberg about 1400.

In 1511 the Teutonic Knights elected as their Grand Master the Margrave Albert of Ansbach and Bayreuth, a Hohenzollern prince and also a kinsman of the King of Poland. It was hoped that his election would aid the knights against their enemy, Poland. However, although the whole country eventually benefited greatly, the almost immediate result was to end the rule of the Teutonic Order. In 1525 the Grand Master became a Protestant and proclaimed himself Duke of Prussia, changing the constitution from an ecclesiastical to a temporal duchy.

Albert proved a just and sensible ruler, and resourceful, as indeed he had to be. When the eastern region of the Teutonic Order's territory became the Duchy of Prussia, Duke Albert took over all the rights belonging to the Order, in particular the lucrative amber monopoly. Unfortunately for the Duke, this was a time of unique and unprecedented sales crisis: the Reformation had virtually killed the demand for rosaries which throughout the Middle Ages had been the sole economic use of amber.

A steady demand and a ready market, with the paternoster-makers of Lübeck and Bruges taking delivery of raw material by the ton, was the secret of the profit which the Order in its prime had derived from amber. Into this centuries-old sales situation the Reformation had remorselessly intervened. The regions turning towards Lutheran doctrine had no further need of rosaries, while the countries that stayed Catholic may well have declined to continue commercial relations with those that had turned Protestant, most particularly the ultra-Protestant Duchy of Prussia.

Thus the paternoster-makers' guilds lost their traditional customers and so their ability to absorb the raw material. New markets and new products had to be found, so that a material which had been used for so long for one purpose could be profitably employed in other ways.

First the cost of production had to be lowered, and Duke Albert arranged this right away. The Teutonic Knights had always paid the shore inhabitants with money and salt; they had plenty of the latter as the Government had that as a

Oak gamesboard, overlaid with amber, tortoiseshell and ivory, German, 1620.

monopoly too. The new administration stopped the money and paid only in salt. This, of course, made life for the unfortunate shore dwellers harder than ever.

The Duke then set about finding purchasers able to pay better prices than the guilds in Lübeck and Bruges. Negotiations were begun and agreements concluded with merchants in Königsberg, Danzig, Lübeck and elsewhere, and again the working of amber by any unauthorised person was prohibited throughout the country.

The Duke reserved the right to employ an amber turner who worked solely for him. He also kept the largest and best pieces of amber for his own use, some of these pieces being as big as a man's head; from these he had many beautiful articles made, both for his own household and for gifts to friends and relatives. So, during the first half of the sixteenth century and utilising the amber no longer required for rosaries, carvers and sculptors were encouraged to produce individual works of art.

Apart from his enlightened patronage of all artists, Duke Albert was constantly seeking new ways to make his amber monopoly sufficiently profitable. Not satisfied with the new measures taken so far, he called for two of the top physicians belonging to the ducal court and set them to work researching into the medicinal uses of amber. This they did, and these two physicians — Aurifaber and Göbel — subsequently produced the earliest scientific treatises on amber, works which show that the Duke and his medical advisers did their utmost to promote the precious substance as an important universal remedy, to be used either externally as an amulet, or internally in the form of a powder or tincture. White amber in particular was used for medicinal purposes.

This novel use of amber became an important way of exploiting the local product. But above all the Duke had decided that the new free Prussia, emerging from the rule of the old Teutonic Order, was going to cultivate its own artistic policies and traditions. During the hard times of the sixteenth century, he gave every encouragement to the arts at Königsberg in the culturally deprived east of Germany. The fame of the local amber workers spread, like their products, far abroad.

Apart from the individual artists, the guilds of amber-turners of course kept going, and new ones arose. The oldest guilds were, as mentioned above, the paternoster-makers of Bruges and Lübeck, both dating from the early fourteenth century and at a safe remove from the point of collection and would-be illicit suppliers. Gradually guilds appeared further east. They were functioning at Stolp*[1] in Pomerania and in Danzig by 1480, and at Kolberg*[2] and Köslin*[3] by the mid-sixteenth century. A guild at Elbing*[4] was founded in 1539 and, by the late sixteenth century, was so firmly established that individual artists working outside it needed official authorisation and protection (see the case of Georg Raport in Chapter 6). However the tradition of not encouraging amber workers near the actual source was so strong that Königsberg — where much of the finest work was individually produced for the courts of Europe — did not have an official amber workers' guild until 1641, by which time its artistic prime was almost over.

The rosary trade eventually picked up again to form the staple work of the guilds. Leaping ahead in time, as late as 1745 the test for a journeyman of the Königsberg amber workers' guild to qualify for the rank of master was to carve by eye a string of round beads suitable for a rosary. Oddly enough the technical term for amber rosary beads was "corals", so the Königsberg document specifies: "One quarter pound of perfectly round corals measured by eye without aid of compasses, wherein the holes must be drilled even and straight." A bill presented by the guild at Stolp mentions "pale olive-coloured corals". On the other hand the guilds did make other objects beside the rosaries. Thus a Lübeck document of 1709 itemises "measuring vessels, cups, boxes, spoons". The anonymous masters would also supply cabinets, crucifixes or mouthpieces for pipes. In at least one case individual names emerge; those of Johann

*[1] now Slupsk, in Poland
*[2] now Kolobrzeg
*[3] now Koszalin
*[4] now Elblag

Segebed and Niklas Steding, both of Lübeck, who between them made a cabinet and two crucifixes.

The products of the Stolp guild were sent as far afield as Constantinople, Smyrna and Aleppo, while the Königsberg workers maintained a regular trade with the Turks and Armenians. Duke Albert's enterprise had certainly proved worthwhile.

From the beginning of the fifteenth century, amber beads often appear as legacies. Sometimes just two beads represented a legatee's sole benefit under a will; "a pair of amber beads" is a frequent form of bequest. Inventories of French legacies and bequests of the fourteenth and fifteenth centuries mention individual carvings in amber. An amber collection that included a particularly fine cup was left to the ducal treasury at Gotha by Elizabeth Sophia, daughter of the Great Elector of Brandenburg.

A late sixteenth century inventory of the bequests of Queen Elizabeth of France mentions "a drinking set of whole amber" which the Duchess of Prussia had once presented to the Queen, and which was valued in the inventory at 60 florins. This exquisite set passed into the possession of the Hapsburgs, but it seems to have disappeared later.

CHAPTER FIVE

THE AMBER GATHERERS

In this chapter, the various ways of collecting the raw amber — a difficult task at all times — are described. When the violent north-west winds and heavy seas had loosened the amber from the sea bed and swept it along with great tangles of seaweed on to the shore, it was a question of good fortune for the coast dwellers whether a gust of wind would drive the amber towards their particular bit of beach. If the winds did not favour them, they must watch the amber being collected by one or other of their neighbours.

There were some, however, who were not content to risk losing the precious bundle. It was possible to scoop up a tangle as it floated upon the surface of the water. A special net was used for this called a *kascher,* which resembled a butterfly net attached to the end of a 20 foot pole. With this net the men waded as far as they could into the sea and lifted out the tangles of weeds and amber borne along by the

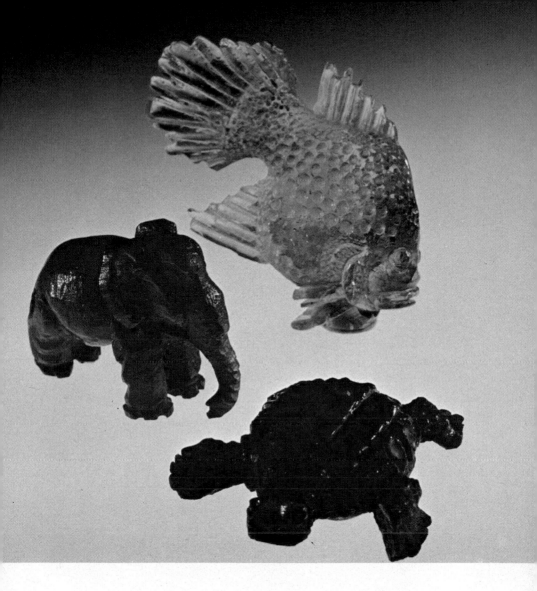

5. Three animal carvings: fish, made in Königsburg, 78 mm long x 51 mm high; elephant, Chinese, 53 mm long; tortoise, Chinese, 53 mm long.

6. *Overleaf:* A fine natural clear piece of amber with 'underwater' inclusions, 62 mm across side shown.

waves. The women and sometimes even the children helped in this work by waiting on the beach, picking up the pieces of amber as they were thrown to them and finally putting them in rough sacks kept for this purpose.

Generally the most violent and productive storms occurred in the months of November and December, so that the amber fishermen had to wait for the right moment. These men had to be strong and vigorous, as the weather by then would be bitterly cold. They also had to wear special protective clothing of leather and wool. At times the weather was so severe that their clothing became completely frozen and had to be thawed out at the fires which were kept going by the women on the beach nearby. When the weather was particularly bad, and the lives of the men were at risk from the huge waves caused by the storm winds, they formed a line connected by a strong rope, after the manner of mountain climbers.

This was the oldest method of gathering amber, and the amounts collected varied greatly. Sometimes in very adverse conditions the yield was disappointing and hardly worth the toil or the risks involved. In other places where large boulders lay close to the beach, a different way had to be found to collect amber. The force of the waves was broken by the stones and the amber fell among them.

With the system called *bernsteinstecken,* the men would put out to sea in small boats to fish the amber up from the bottom. The sea had to be very calm for this as it took extremely sharp eyes to see the amber even when the water was very smooth. One boatman would loosen the pieces with a special kind of spear while another man held out the net ready to catch them. The poles and spears varied from 10 to 30 ft/3 to 9m. in length; the spearhead was made of iron and shaped like a half-moon. Sometimes enormous

SEVENTEENTH-CENTURY AMBER GATHERERS, NORTH PRUSSIA.
Two fishermen with their net and their leather bags
for collecting Amber.

From the first book that was written on the Amber industry,
" Succini Prussica, Physica et Civilis Historia," by P. J. Hartmann,
published in Frankfort in 1677.   A very rare volume.

boulders had to be moved to free the amber, and then large forks with prongs were used to do it. During these operations the boat would tip so far over that the gunnel came down almost to the surface of the water.

Yet another way of collecting the precious stone was called *steckerei.* This was used when a rich deposit of amber lay deep down under the water and extended east and west along the shore. Great quantities of the large stones had to be brought to the surface from the particular area and carried away on rafts. The sea bed was then swept with nets provided with sharp rims to dislodge the small stones and the lumps of amber. Round about the sea would be dotted with small boats, all the men intent on bringing as many pieces up as possible.

A later and more costly method was dredging, using both hand and steam dredgers. Hundreds of men had to be employed to carry out the work, and it also needed the co-operation of business men who undertook to keep the channels open and who paid out considerable sums for the privilege and the right of obtaining the amber.

In the seventeenth century the first amber mines were investigated and worked. The earliest systematic mining began in the brown coal sands of the Samland peninsula, which had rich amber deposits. Later on, about the middle of the same century and again in Samland, mining began in the famous "blue earth" area, and this became an important source of amber. When the value of the seams was recognised, capital for the working of the mines was soon forthcoming.

Huge pits were dug in the sides of the cliffs, and as the earth was taken out it was piled into crates and conveyed by tramways to the beach. Quite often operations became difficult because of the influx of water as the mines descended as much as 40ft/12m. below sea level. Each workman, equipped with a special spade, dug slowly and deeply into the earth until the spade came into contact with a lump of amber. This was very carefully freed from the surrounding earth. The cleaned amber was wrapped in large coarse sacks kept close by. Men were employed in these mines all the year round, and overseers were stationed near them while they worked to see that they did not steal any of the precious pieces.

A detailed description of the procedure followed in working such a mine was given by the writer Felix Dahn (1834-1912): "The workmen stand in three parallel rows, knocking to pieces every clod of the blue earth, in which the amber was found. A group of six or eight men were placed in the charge of each overseer. While he watched the clods of earth were first thrown into a large vessel of water and carefully examined by a few of the men carrying out this part of the operation. If they found no amber at all in the lumps of clay it was thrown from the lowest level up to a second platform which could be reached by long ladders. Here the refuse material was taken in charge by another group of men and women and it was again flung from shovels to the highest platform of all and then carried away.

"While the work was being done the overseers kept up a most monotonous rhythmic singing accompanying the regularity of the movements and apparently this was all part of a deliberate policy in order to prevent pilfering. This was not always successful, however, even though the miners were searched before leaving the pit after the day's work."

What a strange and intriguing sight this must have been. Men, women and children, some in the oddest attire and all of them exposed to the sharp cold winds, but all digging vigorously to the languid sounds of the sombre melody.

Dredging for amber near Memel (now Klaipeda), Lithuania.

Prussia in those far-off days produced about 220,000 pounds of amber each year, out of a total of 250,000 pounds for the whole world. Most of the pieces found did not exceed one pound in weight, with a few very special exceptions. In 1803 a piece was found which weighed 13½ lb/6 kg, and this rich prize was placed in the Royal mineral cabinet in Berlin. It was also recorded in 1681 that one of the Dukes of Brandenburg had presented the Tsar of Russia with a chair made entirely of amber. Another writer speaks of a single lump so large that a man could hide behind it, which was found in 1596. He added that the piece had to be broken up as no one could be found rich enough to buy it.

CHAPTER SIX

ARTISTS IN AMBER: THE LATE 16TH AND EARLY 17TH CENTURIES

I have seen amber big as men's heads wherefrom his Serene Highness in Prussia had beakers and bowls made", wrote Andreas Aurifaber at the time of Duke Albert of Prussia. I have shown in Chapter 4 (Amber in the Middle Ages and Renaissance) how the Duke, both for economic and cultural reasons, came to encourage individual artists in amber, ordering sculptors to use the material in new ways, and reserving the finest quality and the largest pieces found for this purpose.

In 1563 the ducal court engaged its first specifically artistic amber carver. His name was Stentzel Schmitt, and he was to receive a salary of 100 marks per annum, plus the bonus of one court robe a year. He was to undertake all the Duke's amber carving, "be it large pieces or small". In 1567 the old court accounts show him commissioned to make two fine amber chessboards representing hunting scenes.

Records of his artistic activities can be found for another 25 years, for he continued as the court amber carver under Albert's successor, the regent George Frederick of Ansbach (Albert's son, Albert Frederick, was mentally deranged). George Frederick obviously took to this exotic material, new to the stranger from Franconia, with real enthusiasm. When he left Königsberg in 1586 to return to Ansbach, he maintained his connection with the East Prussian amber craftsmen. In the following year, he sent for Stentzel Schmitt to come to Franconia for several months.

A few years later one Hans Klingenberger, "amber turner of the Rossgarten", was paid 600 marks for an amber box, two carving knives, seven whips with amber handles, and a quantity of other amber-handled cutlery, "artistically fashioned", which were sent to George Frederick at Ansbach. By then Stentzel Schmitt was presumably dead, or no longer working.

By 1959 Klingenberger was so busy that he was forced to delegate some of the work to one Georg Raport of Elbing. The Prussian government wrote to the Elbing town council: "Our amber turner Hans Klingenberger has been compelled to transfer some of the work entrusted to him to one of your craftsmen, Georg Raport. We request that you do not allow your amber turners' guild to prevent Raport from completing the work or to obstruct him in any way." One hopes that in the event Raport was not victimised.

Other individual craftsmen are named in the court records: Michel Meier, Kasimir Zweck and Michael Fischer. Joachim Schönemann was one of the most famous amber carvers of the early seventeenth century. Outside Königsberg, there was Peter Hegenwald of Danzig, and Daniel Hindenberg of Lübeck.

Of the many objects produced by these amber carvers,

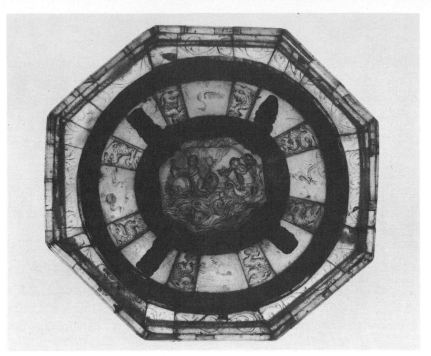

German cover and bowl (opposite) with marine reliefs.

the following are specifically mentioned in the old accounts: compasses, powder flasks, inkwells, dice, hunting horns, turned and carved beakers and cups, clock cases, chests, bowls, jugs, jars, spoons, carving knives, walking sticks and whips, as well as necklaces, bracelets, pendants, religious statuettes and portrait medallions.

Often a setting of gold or silver is mentioned in the accounts, especially in the case of jugs, tankards or sweet dishes. Whole sets of church vessels were made: chalices, ewers, water sprinklers, candlesticks, crucifixes, altars for private chapels — even bureaus and small cabinets. The glowing light and warmth of amber gave it a special appeal for a religious age.

Only a very small percentage of these beautiful works of art have survived the ravages of time and the devastating onslaught of modern warfare. Those that remain, scattered

They measure 310mm across × 125mm high, mid 17th century.

over the world in museums or private collections, are priceless treasures. The astonishing diversity of objects shows how deliberately amber was used as an artistic export, and how, by means of diplomatic gifts, the Prussians sought to enlist the admiration and desire of foreign rulers for their local product.

Among the Danish royal family's art treasures at Rosenborg Castle was a set of 18 large plates with broad silver rims and translucent amber centres; on the underside of each plate the coat-of-arms of Brandenburg and Brunswick-Lüneburg, worked in ivory or white paste, was inlaid into the amber, together with the date 1585 and the initials of George Frederick of Ansbach's second wife Sophia, daughter of Duke William of Lüneburg. George Frederick did not leave Königsberg for Ansbach until 1586, and it seems most likely that the amber centres of these plates, whose beauty was marked by an elegant and functional simplicity, were the work of Stentzel Schmitt and his workshop.

The increasingly ornate cups, chessboards and knife handles of the late sixteenth and early seventeenth centuries show a great use of ivory and different coloured ambers together, ivory portrait medallions on vessels of light and dark amber, or amber plaques set into ivory knife handles. As well as carving, there were engraved or incised decorations, and coats-of-arms applied by the colourful verre églomisée technique.

The finest work of the first half of the seventeenth century was done at Königsberg, and the greatest master of that period was Georg Schreiber, who was active until about 1643 and who, unlike earlier craftsmen, tended to sign his work. Normally he signed plain "Georg Schreiber", but in the case of one magnificently carved tankard mounted with silver, he signed proudly in Latin: *"Georgius Scriba Borussus Civis et incola Regiomonti Borussorum hoc fecit 1617"*. (Georg Schreiber, Prussian citizen and inhabitant of the province of Prussia, made this in 1617). His splendid amber chests — some of them in several tiers — were not built up on a wooden core, but consisted of innumerable panels of richly carved amber fitted and glued together, or possibly hinged with metal screws; sometimes mounted with silver or silver-gilt, sometimes standing on ivory base or feet. The decorative carving, which towards the end of the period covered every available inch of space, might be of animals or plants, or scenes from classical literature and mythology, The Judgment of Paris was a favourite.

Another artist of the mid-seventeenth century was Jacob Heise who produced, among other works, several signed cups in the shape of nautilus-shells, exquisitely carved and

Judgement of Paris, possibly by Gottfried Leygebe, 197mm, 17th century.
(See page 52)

Casket with bone reliefs in sarcophagus shape, 235mm long × 260mm high, late 17th century.

crowned with the figure of a man riding a sea monster. A low chest, surmounted with two standing figures which are realistic portraits of the Great Elector Frederick William of Brandenburg-Prussia and his wife Luise Henriette in full robes and regalia, was probably also the work of Heise. The drawers of the chest were carved with hunting scenes and cupids riding on sea monsters, similar to Heise's nautilus cups, and the chest also had inset medallions of carved ivory under a transparent amber overlay. The figures themselves were not carved from solid blocks of amber, but were hollow and built up of a patchwork of clear and opaque ambers, with possibly some use of meerschaum. This interest in realistic detail, free-standing sculpture and the patchwork of different ambers that made such intricate treatment possible, seem to have been characteristic of Heise's art. The Great Elector was an enthusiastic patron and promoter of amber work, and even tried his own hand at it, we do not know how successfully.

By the mid-seventeenth century, Königsberg had begun to decline as an artistic centre, and was no longer the source of the finest amber work. The focus had shifted to Danzig, where many talented artists were active. In Danzig as in Königsberg there was a powerful guild of amber turners, but as ever the outstanding works of art seem to have been produced by individuals, who sometimes were not even members of the guilds. Probably the output of the guilds continued to be mainly rosaries, even during the great period of the seventeenth and early eighteenth century, a phenomenon that is described in the next chapter.

It is a revelation to look at what is in the museums. Of course there are larger collections abroad in Germany, Scandinavia, the Baltic countries and at the Pitti Palace in Florence; but examples of the finest work can also be found in London. At the Victoria and Albert Museum, there are two representations of the Judgment of Paris which must be ranked among the finest of amber sculptures. The earlier of these is a plaque carved from one roundish piece of amber roughly 5 inches/13 cm in diameter. The scene is beautifully composed to fill the entire space: to the left Paris sits on raised ground, holding the apple of discord in his right hand. Venus, wearing a draped loincloth and a necklace, has put her right arm about his shoulders in a charming gesture of confidence (her left hand is holding her own breast). To her right are the two rejected goddesses, both fully dressed; Juno with her peacock, and Minerva with her owl and wearing a helmet as always. Cupid, with a pet dog, sits in the foreground. Treetops finish off the top of the carving. An inscription on the smooth back reads "Carolus Maruti ex. 1621". This piece is believed to come from Danzig (though one German authority, Alfred Rohde, ascribed it to an unknown Königsberg master working later in the seventeenth century).

The second Judgment of Paris has been treated as a freestanding sculpture. Here, too, the main group would seem to have been carved from one great find of amber, though the base is composed of plaques of multi-coloured clear and opaque ambers. Paris and Venus sit facing each other, his left hand and her right hand holding the apple, which forms the centre point of the sculpture. In front Cupid is playing with his quiver full of arrows. Above them looms the trio of Mercury with his winged helmet, Juno and her peacock on his right, and Minerva with helmet and owl on his left. The exquisitely intricate group is rounded off with flowers and grass, and cherubs decorate the base. This is late seventeenth century work, from Danzig.

There are also two splendid amber and ivory caskets,

Judgement of Paris, signed 'Carolus MarVTI Ext 1621', German, 133mm high.

dating from the seventeenth century. The more elaborate of these, a two-door cabinet on six amber feet, is a sort of miniature palace. The doors open on an archway and an inner compartment with three pillars and a mirror at the back, and a floor mosaic of amber and ivory. The archway is flanked by five small drawers on each side. Such an intricate piece shows the full development of the incrustation technique.

Among other works of the great period are an amber shrine decorated with ivory plaques and figures, and an altarpiece which Rohde considered the most magnificent and artistically mature work of its kind. The base shows Christ and the apostles; the middle section is dominated by an amber relief representing the adoration of the shepherds, flanked to right and left by the figures of the Virgin Mary and the Angel of the Annunciation; the upper portion is ornamented with three large ivory plaques of scenes from the life of Christ. The reverse side is carved with equal richness and elaboration, the centre showing a perpetual calendar with the signs of the Zodiac and above these the Last Judgment.

This by no means exhausts the list of amber treasures to be found at the Victoria and Albert Museum. I will mention only two more: a Sicilian carving of the Holy Family, with amber figures unusually set on a background of lapis lazuli; and a delightfull small sculpture of a resting lion. This is an earlier piece, from the sixteenth century, very simple and with a rougher finish, but full of spirit.

There is less amber on show at the British Museum, but two pieces worth searching for are a casket set with engraved amber (c. 1600) in the Waddesdon Bequest, and a magnificent amber tankard with cover, made up of eight vertical panels carved with figures of the cardinal virtues

7. A rare faceted and graduated bead necklace from the Imperial Russian court, approximately 130 cm long.

8. *Overleaf:* Turkish pipe in its original case, carved from a solid block of amber, with the bowl shaped as a man's head. The head is 57 mm high.

Amber and bone candlesticks, German, 305mm high, c. 1700.

and dated 1659 on one panel. This tankard is mounted in silver-gilt and the handle bears a sort of amber mermaid like a ship's figurehead. It probably came from Königsberg, a perfect combination of glowing natural beauty and the finest craftsmanship.

So at least some of the marvellous work of past artists in amber remains, to show what could and can be done.

CHAPTER SEVEN

AMBER IN THE BAROQUE PERIOD

Before the 1930s, a great deal of the artistic work in amber was based on the technique of incrustation. This method, a baroque style of extraneous decoration superimposed on a second and necessarily plain material, was used to produce many fine treasures. It was particularly effective in the making of snuff boxes, jewel caskets, cabinets, altar-pieces, chessboards and many other things to which this type of work was applicable.

As early as the sixteenth century, there were many scientific books written on the subject of amber, but few, if any, on its uses in the world of art. Unfortunately the extensive use of the incrustation technique gave a mistaken impression that this was the only type of work produced. Following this erroneous conclusion, it was logical that amber was looked upon as a somewhat fragile material and the end products ephemeral.

The first book to be written on the artistic history of amber was published in 1920. Its author was Otto Pelka; he made a serious effort, in what was after all a pioneer work, to show the many ways in which artists and craftsmen through the ages had used this marvellous material. However he was hopelessly handicapped by the inferior techniques of book production, layout and pictorial reproduction of his time; the illustrations totally failed to convey the splendour of the works. His modest and, it must be admitted, unattractive-looking volume could hardly change the prevailing idea of amber as a material suitable only for trinkets and souvenirs. Yet all the time evidence existed of amber used on the most heroic scale.

The most magnificent example of the incrustation technique the world has ever seen was "The Amber Room". This unique work of art was begun at the request of King Frederick I of Prussia, very early in the eighteenth century. An expert Danish amber carver and turner, Gottfried Wolffram, was recommended to Frederick by the King of Denmark, and it was this man, together with an architect, Eosander von Göthe, began the construction of an amber room at Charlottenburg.

Working continuously for six years, Wolffram completed, panel by panel, a whole wall. At this point he seems to have had a serious disagreement with the architect which resulted in his dismissal, and though he endeavoured to obtain compensation for loss of employment and breach of contract, he did not succeed.

In 1707 the work was handed over to two other craftsmen from Danzig, Ernst Schacht and Gottfried Turau. Between them they finished the panelling by the end of 1711. There seems some doubt, in fact, as to whether the room was ever assembled at Charlottenburg, but in 1713 it was set out in all its glory in the royal palace at Berlin, and it was here that the Tsar of Russia, Peter the Great, saw this incredible work of art. He was so impressed with the room that the Prussian King insisted that he accept it as a gift. The Tsar, unable to resist, had the room dismantled and very carefully packed into special boxes, after which it was taken by sleigh to his winter palace at St Petersburg. Evidently Peter the Great intended to incorporate the room in the new palace buildings that he was in the process of designing.

In 1755 the Empress Elizabeth, daughter of Peter the Great, commissioned the sculptor and amber carver Martini to instal a room at the Winter Palace using the amber ornamentation. But once more the work was delayed as the Empress decided to have all the materials, together with the finished panels, transferred to the Summer Palace at Tsarskoe Selo, outside St Petersburg. This was done and the room was almost completed; though as late as 1763, various craftsmen were still putting finishing touches and making small luxurious additions to the now prodigious creation.

One privileged Russian viewer described the final effect of the Amber Room as follows:

"The room is certainly one of the Wonders of the World. Upon entering it the visitor is overwhelmed by a sense of awe and a feeling that he has stepped into a kind of dream or fantasy. It was, however, no fantasy, but a creation of such beauty that one could hardly believe it real."

The character of the room was mainly baroque. The walls were covered with a polished mosaic of amber pieces of irregular sizes, all in varying shades of yellow. Finely carved frames divided the mosaic into panels, and in the centre of the panels, Roman landscapes represented allegories of four of the five human senses. In addition there were carvings of

Front of an amber and ivory altarpiece, German, 17th century.

Rear of the same altarpiece, featuring pierced ivory plaques of the zodiac.

many kinds: flowers, shells, garlands and trees, with curved amber plates at intervals which were inlaid with tiny figures carved with incredible precision. These figures were so small that it would have needed a magnifying glass to view them with any success. Here and there shiny plaques of clear red amber glowed in brilliant contrast to the mass of yellows and browns which made up the whole. The crucial dates in the history of the room — 1709 and 1760 — were incorporated in the mosaic of one wall.

At each end of the room shone enormous mirrors in gilt frames and in between these were matching chandeliers twinkling with hundreds of amber drops which were cut in facets to catch the light. Doors and windows were also framed in amber which was the colour of deep gold. Ornate showcases stood around the floor filled with small but valuable amber objects. The room was lit from both sides, and the effect was one of inexpressible charm. The whole decor was harmonious and pleasing, with nothing ostentatious or gaudy about it. The effect of the amber wall panelling was not unlike that of marble but, in complete contrast to the feeling of chilly magnificence that one associates with marble, it seemed to exude a special warmth, while its beauty far exceeded the most sumptuous wood panelling.

Unfortunately the Amber Room was not to remain in the possession of the Russian people. In 1941 when the Germans were closing in on the city of Leningrad, it became vitally important to move as many of the priceless art treasures as possible to underground vaults at Sverdlovsk in the Ural Mountains. Before this could be accomplished, the contents of the Tsarskoe Selo palace had vanished, leaving only an empty shell, and with them went the Amber Room.

Some five or six years later the Soviet government began

to look for their lost treasures, and in particular for any signs of the precious room. According to a German officer who had been taken prisoner in 1944, orders had been given by the German High Command to dismantle the room and transport it to the Prussian Fine Arts Museum in Königsberg under the care of that museum's director, Dr Alfred Rohde, who had the room assembled in Königsberg Castle.

By the end of 1943, plunder from Russia and elsewhere filled the castle so that it became impossible to accommodate anything more. Intensive bombing attacks by the British were taking place regularly, some bombs fell uncomfortably close to the castle, and Rohde began to fear for the safety of the priceless treasures for which he had been made responsible, most especially the Amber Room. Once again the room was dismantled and stored, among other precious things, in an underground cellar beneath the Castle. But apparently Rohde was still not satisfied with this place of concealment. When the Russians reached Königsberg in April 1945, all trace of the Amber Room had disappeared; as had Alfred Rohde.

However Rohde returned and took up his former position, now under the Russian occupation. He appeared co-operative and helpful, but was adamant in denying all knowledge of the Amber Room. Early in December he fell ill, and on 16 December both he and his wife died from a type of dysentry. The death certificates were signed by a Dr Paul Erdmann, but this mysterious doctor could never be traced. Nor has anyone since discovered Rohde's secret.

There has been some speculation regarding the attempt in 1972 to salvage a sunken ship from the bottom of the sea some 20 nautical miles off the Baltic coast of Germany. It is just possible that this vessel, which had been torpedoed by a

Bowl with figures in relief, German, 210mm × 76mm high, c. 1650.

Soviet submarine in 1945, could contain the marvellous Amber Room. Though the first attempt at salvage failed, owing to tempestuous weather, another is to be made in the near future. If the next attempt were to meet with success, and the room be recovered without too much damage, it will be handed back to Russia, its rightful owner, and set up once more where it will be the proud privilege of millions of people to view this priceless work of art in all its marvellous beauty.

Individual patrons, even as enthusiastic as the Great Elector, cannot by themselves explain the sudden upsurge of an art form. But certainly the late seventeenth and early eighteenth centuries were the great age of the German provincial courts as patrons. The numerous landgraves, dukes and princes vied with each other in patronising the arts and furthering culture and enlightenment. It was not only Königsberg and Danzig, every provincial court

Top: Gamesboard decorated with bone plaques, German, 735mm × 370mm, 17th century.

Above: The same gamesboard closed showing the chessboard and part of an amber chess set, early 18th century. The other side of the gamesboard has a board for Nine Men's Morris.

employed its own amber carvers and sculptors. The amber workers of this period had all the skills: they were equally at home with the lathe, with carving and engraving.

The very difficulties presented by the material appealed to the Age of the Baroque, with its passion for scientific curiosities and technical ingenuity. Amber was used in all sorts of unprecedented ways. For instance, in 1691 Christian Porschin, a member of the Königsberg guild, made burning-glasses of amber, which were said to be much more effective than glass lenses. He also ground lenses for spectacles out of amber, and polished prisms, "using a special process of gentle heating in linseed oil to render the amber clear and transparent". There were egg-timers made of amber. Frederick the Great of Prussia, who among his other numerous activities loved to play the flute, was on one occasion presented with a flute made of amber.

Indeed, there were many princely presents made of amber. Just as the age appreciated the scientific strangeness of the substance, so they revelled in its decorative qualities and the chances it offered of lavish ornamentation. Exquisitely carved chests, boxes, chessboards, sets of cutlery and drinking vessels of every shape and form were exchanged; some of them can still be admired in museums and art collections all over the world. Frederick the Great, recipient of the flute, in his turn presented the Empress Elizabeth of Russia with a mirror frame of amber which was later incorporated in the Amber Room at Tsarskoe Selo. In 1687, Louis XIV presented the Siamese ambassadors to the French Court with several amber boxes and two mirrors with frames of variously-coloured amber.

The amber turner Michael Redlin of Danzig, (typically enough, the surviving records of the local guild do not list him as a member), made three magnificent objects which were then given by the Elector Frederick III of Prussia to the joint Tsars Ivan and Peter. These were a two-tier amber and ivory cabinet carved and engraved with landscapes and historical scenes, flowers and foliage; a chessboard of multi-coloured ambers, complete with matching chessmen; and a twelve-branched chandelier of ambers with ivory, inlaid with metal plaques painted with portraits of Roman and German emperors and heroes under a clear amber overlay. No wonder that the master swore that he had spent two years working on this chandelier alone. The total cost of all this, including the packing which must have taken some care, came to 2,282 florins. Unfortunately the objects themselves do not appear to have survived the various fires and disasters of Moscow, but the artist's hand-drawn designs did.

Another independent craftsman, Christoph Maucher, who was born in South Germany but worked in Danzig in the late seventeenth century, aroused the anger of the guild because he was given commissions by the town council of Danzig as well as by the Court at Berlin. The guild accused him of "profiting by other men's work, though he was neither a citizen nor a fully qualified guildsman". However their attacks were in vain, since his work "was held in such esteem that in all Europe there was not his equal".

The late seventeenth century produced some of the finest amber sculptures known. They were works of pure sculpture, not merely applied decoration. Among them were portrait medallions (there was a particularly good one of Frederick III himself), domestic altars richly carved with scenes from the Bible, reliefs and free-standing figures and groups, the last of either religious or classical and mythological subjects.

Towards the end of the century, the use of ivory with

Two reliquary pendants set in gold. Left: late 17th century; Right: 18th century.

amber was almost entirely ceased. The two materials were used for separate sculptures. The two materials were used for separate sculptures. The decorative variety of colouring was achieved by mosaics of clear and cloudy ambers in all possible colours, and sometimes including the use of dyes. There was also an increasing use of the technique of incrustation; the cabinets and crucifixes were no longer made simply of panels of amber, but of thinner panels plus inlay on a wooden core.

The most famous of the later sculptors was Jacob Dobbermann, who was employed by the court of Cassel in the early eighteenth century. His skill and confidence completely overcame the fragility of the material while giving full value to its gorgeous colour and texture. He used many classical themes, including the Rape of the Sabine Women, Saturn and Truth, a Diana and a Cleopatra. The boldness of his design and perfection of his execution were astonishing. He also produced works of pure craftsman-ship, such as an eight-branched chandelier, but he was obviously less interested in such commissions; apart from the rarity value of objects of such size, the chandelier seems to have been a rather inferior piece which disappointed his patron, the Landgrave Karl of Hesse-Cassel. Some lovely shell-form cups — the insides carved with groups of lovers, cupids or nereids, the rims and stems and bases adorned with sea-shells, foliage and fauns — which were also created for the court of Hesse-Cassel, have been attributed to Dobberman's contemporary, Johann Caspar Labhard.

Yet even at Hesse-Cassel, things were changing. When Landgrave Karl died in 1730, the court seriously debated whether to keep Dobbermann on at all. "He is certainly very skilful, but his work is of no practical use," wrote one official bluntly. It is to be hoped the vote went in his favour; at any rate, when Dobbermann died in 1745 he was still described as "Court artist in amber".

From this period onwards, the importance of amber as a medium for artists began to decline. Eighteenth century society — this was the Age of Reason — had little of the religious imagination or the romantic feeling for natural beauty which had inspired the artists and patrons of the preceding age. In 1742 the amber workers of Königsberg petitioned Frederick the Great, complaining that as no one would buy their work, that perhaps he would help them by his patronage and by giving amber presents wherever possible; "seeing that our forefathers fashioned the most precious pieces from this treasure of the sea, the which were sent in great numbers to Russia and other Courts, and admired there as wonders of Nature".

As they lost confidence that their finer skills were wanted, so the artists became imitative and timid, producing little but trinkets, jewel boxes and umbrella handles. The very technique of incrustation, which had seemed such an advance and had produced such brilliant things, now contributed to the decline. It seemed to stress the fragile and superficial qualities of amber compared to the solid core of wood. Along the whole Baltic coast only the craftsmen of Stolp were still busy, and their main output was necklaces for peasant women.

However, there were still a few artists like F.W. Arnold, who in 1821 carved an allegorical figure of Hope on the lid of an amber snuffbox, and inscribed it, "The guild of amber workers in Stolp". Lorenz Spengler, who worked in Copenhagen during the late eighteenth century, produced some fine portrait medallions and at least one very splendid chandelier for the Rosenborg Palace.

CHAPTER EIGHT

THE 19TH AND 20TH CENTURIES

On the whole the artistic content of nineteenth century amber work is low. Refinements of technique, with very few exceptions, disappeared. More and more it became a matter of producing cut or turned amber *en masse,* without any attempt at artistic treatment, but considering only the material value of the stone. The ostentation of the period showed the increase of wealth created by trade and industry without a corresponding growth of taste and discrimination. It was a time for massive brooches, bracelets and necklaces, cigarette holders, mouthpieces for pipes and paper knives. Amber was even used for sewing and crocheting implements. And of course there were all the souvenirs and inlaid or incrusted boxes brought back from holidays beside the Baltic or North Sea. The age was characterised by poverty of artistic invention and a great deal of routine derivative work, and not until the turn of the century were

9. White Baltic and red Chinese amber jewellery. All made in the 19th century, except the large silver-mounted pendant (103 mm across) which is modern.

10. *Overleaf:* Sicilian amber male and female sculpture, made in the 19th century, 76 mm high.

there any hopeful signs of improvement.

In the early 1900s, a group of Danish craftsmen once again drew attention to and revived interest in "the gold of the North". Then the aesthetic ideals of the *Art Nouveau* (or *Jugendstil*) movement which embraced a love for rich colours, flowing lines of decoration, semi-precious stones and individual craftsmanship, plus a new awareness of natural beauty, influenced artists everywhere and helped to bring amber work back into fashion.

Erich Stumpf of Danzig was the most notable amber craftsman of the early twentieth century. His workshop created a wide range of fine decorative objects and jewellery making full use of the colours and qualities of amber. He often combined it with settings of precious metals and other stones; chiefly turquoises, moonstones, amethysts, chrysoprase and cornelian. The simple, elegant forms of his goblets and beakers, bottles, boxes and paperweights were the perfect complement for the sensuous richness of his favourite material. A lesser craftsman, who did not specialise in amber but still produced some beautiful work, was Paul Pfeiffer of Pforzheim.

The modern trend was more and more to take pleasure in the magnificence of surface, colour and texture of the stone "in its natural beauty", either alone or in combination with other precious materials, and there was little impulse to go in for artistic carving, engraving or sculpture.

The amount of amber produced today bears little comparison with that of earlier centuries, though it is still from the coasts surrounding the Baltic Sea that the greatest proportion is procured.

One recent and exciting discovery of a considerable quantity of amber was made in the neighbourhood of Gdansk, formerly the City of Danzig, in Poland, during

A set of six yellow napkin rings, 33mm across, 19th century.

excavations connected with plans for the extension of the port of Gdansk. The men who were chiefly responsible for unearthing this treasure trove were two Polish engineers who were also twin brothers. It is relevant at this point to stress that these were amber men living in an amber land and that the discovery was, to them, comparable to striking oil: it meant wealth and a chance of success for the future.

It all began with the need to reinforce the ground where the work was to take place by means of concrete piles. One of the brothers suggested that it might be more effective if, instead of drilling deep down into the earth in order to sink the piles, they used a large hose pipe under high pressure to wash the earth out. The idea was put into operation right away, and the totally unexpected happened; the hose began washing out great lumps of amber.

The excited engineers could hardly believe their eyes, but there was no doubting the reality of the beautiful golden fossils in their hands. They immediately started to make plans to instal a pump and get it working, in the hope of finding a haul of amber big enough to justify their joint expenditure. It was a chance they were both prepared to take, even though it meant giving up their employment with the company they were working for. They decided to go ahead. The first difficulty they came up against was where in the district to obtain the water necessary for the working of the pump. However, they were able to use the irrigation system of the neighbouring fields.

Good fortune now seemed to favour the enterprising brothers. Right at the start they encountered a great cluster nest of amber, and it was rumoured that for the next two weeks the pump was washing out 440 lb/200 kg of amber a day. By then thirty more men had joined the venture, each contributing another pump. The excitement was intense: they were gambling, and the prize was amber. Each morning the prospectors gathered round while the enormous hoses penetrated the ground. The pressure was applied, and there was silence as they waited to see what came up. Had they hit more or not? From morning till night these gamblers were soaking wet, splashed with mud and stinking sewage, but it was worth it! They drew 30 tons/30,480 kg out of one cluster nest alone.

Then trouble began, for the pumps had destroyed the irrigation installations and the local authorities intervened. Many pumps were confiscated and patrols were instructed to watch the area. The prospectors were upset but they were not going to give up so easily. They could lose their pumps if they were caught hauling up the valuable catch, but if they were lucky enough to get away with it undiscovered they could collect up to 220 lb/100 kg each night. Now they conspired to carry on in secret under cover of darkness, posting lookouts at strategic points while the work went on, and keeping the pumps hidden in old bunkers during the day. This game of hide-and-seek went on for two years until the collection of the amber was taken over by the Committee of the National Council of Gdansk.

The Council decided to grant to private individuals licences valid for one year which gave them the right to pump out amber from the terrain designed for the future extension of the port of Gdansk. The concessionaires paid three million zlotys per year in taxes, and the amber had to be sold officially to craftsmen working for the chain of shops known as "Cepelia" throughout Poland and elsewhere.

This newly discovered method of washing out amber from the bowels of the earth was like manna from heaven to the people who earned a living cutting and carving it, turning it into charming articles of jewellery to be sold at home and

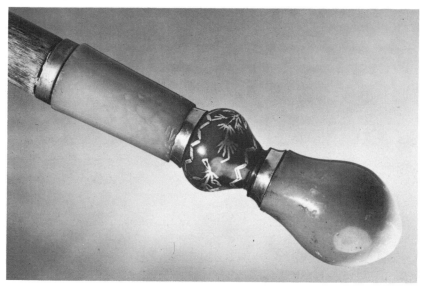

Orange amber handle to a parasol, 152mm long, 19th century.

abroad. Before this, raw material had had to be imported from Russia and, though the Russians sent what they could for the sake of neighbourly relations, they do not have enough themselves to continue to supply it indefinitely.

The Council of Gdansk intended to use the period of licensed private exploitation to discuss plans for State exploitation of the amber fields. But this was no simple matter, and a great many legal difficulties had to be overcome. Officially there are no amber fields in Poland and the mining laws do not apply to amber. When the geologists do find it, it is regarded as a freak, and industrial mining has never been considered. Extraction by the pump method had been studied but rejected for the reason that only 10-15% of the amber would be recovered while the remainder stayed in the ground.

At a meeting of experts from the State Geology Office in Gdansk, the majority were sceptical about the whole business. A local man then suggested that a group of them accompany him to the fields which were being worked by the private prospectors. The experts did this, masquerading as visitors hoping to pick up any small pieces of amber that the professionals had missed. After wandering round for an hour, most of them had pockets filled with amber, concrete evidence that the amber field existed. Immediately after this, the State Geology Office ordered a detailed survey of the area.

Later a television report was presented on the pump method of producing amber. This programme brought together representatives of the State enterprise "Hydrohop" (Water-mining) and the Committee of the National Council of Gdansk to discuss the possibility. It transpired that this same method was already in use for mining phosphates, a process with which they were all familiar, and their experience should therefore make the amber project quite feasible.

Finance proved the next stumbling block. Two hundred thousand zlotys would have to be invested, and these costs were to be borne by the State-run chain of "Cepelia" shops, who would in return receive the yield from the amber fields. The response from "Cepelia" was far from enthusiastic. However, while the official bodies hesitated, the licensed private pumpers went ahead getting the precious stuff out of the ground, to the craftsmen, and to "Cepelia". The ultimate size and organisation of the harvest from this source may be still uncertain, but in the past amber from the Baltic has travelled all over the world, and it is to be hoped that it will continue to do so.

Another most interesting discovery was made some thirty years ago, though this was not comparable to the Baltic find in quantity. A rich vein of a most unusual amber was found

Modern Polish vase, made of fragments of amber cast in resin, about 220mm high.

in the hills of the Dominican Republic. It was of a rather soft consistency, in colours of gold and brown with a darker flecking throughout. Very many pieces were found to contain fine specimens of insects. In 1973 a Miami designer introduced her collection of this modern amber jewellery to the top stores in New York, where it proved immensely popular.

Modern women are very attracted to amber jewellery, and indeed it has never lost its appeal since the Stone Age when women wore it in all its natural beauty and charm. Augmented by today's skilful designing, it fascinates by its combination of today and yesterday, and gives special pleasure to people with a taste for the unusual and exotic.

Amber lends itself to the making of rings in a great variety of designs, both modern and antique. The amber itself is of course always antique by virtue of its origin, but

Three modern rings: 3 pieces, 1 piece and a turquoise, 1 piece; all set in silver.

the gold or silver mounting can be done in a modern style. This could be a very practical, as well as distinctive, choice for young people buying an engagement ring, when the more expensive, hard, stones are out of reach.

Every kind of jewellery has been designed and made in amber, though the quantity is always limited according to the supply of raw material. Within these limits, rare and individual, there are contemporary bracelets, pendants, brooches, cuff links and earrings. With such a wide choice, it is a nice thought that the tenth wedding anniversary has been designated the amber wedding.

Quite recently archaeologists in Lithuania have found twenty-four pieces of amber containing flower petals, insects, and the imprint of fern leaves, which the resin has preserved for possibly 80 million years. This prehistoric treasure was discovered during excavations near the fishermen's settlement of Ventoja and was reported by the Moscow News Agency "Tass" on 5 July 1976. So, though modern finds of amber are few and far between, the story of this eternally fascinating substance still continues.

CHAPTER NINE

RARE AMBER

Some of the rarest ambers in the world are those which come from Sicily, Rumania and Burma. The amber produced by these countries is of such distinctive character and quality that it deserves more particular attention than a fleeting reference to its existence.

The mention of Sicilian amber brings to mind a gentleman who lived in Taormina in Sicily. He kept a shop there and came to London from time to time to sell us some of his beautiful amber. On one of his visits he came to dinner at our house and I was deeply impressed by the colourful clothes he wore, a wide-brimmed black hat and a flowing black velvet cape lined with scarlet satin. I was also impressed by the gift he had brought for my mother, a Sicilian amber necklace. The necklace consisted of sixteen pendants hanging from a thick gold chain, graduated in size and separated from each other by a linked design of delicate gold pieces. Each pendant was a different colour: flashing gems of Sicilian amber in red, violet, orange, yellow and blue. As he handed it over, I exclaimed in wonder that I could not believe that this was amber, but our Sicilian friend assured me that it was indeed amber, of the finest quality to be found on the coast of Sicily.

It is difficult to describe the characteristics of Sicilian amber, but the word that best sums up its charm and fascination is "fluorescence". There is usually a light blue-green translucence on the outside which is not due to a distinct surface layer of a special colour but to the power the substance has of modifying the light which falls upon it. As in the necklace I have described above, pieces may iridesce with a variety of brilliant hues, while the actual colour underneath is a straw yellow or a faint olive green. It is possible that this extraordinary range of colours has some

'Tan-tan' or black amber snuff bottles, Chinese, 78mm and 54mm.

Mottled amber and golden-brown snuff bottles, Chinese, 59mm and 51mm.

connection with its being found in the volcanic earth near to Mount Etna; however pieces have been picked up close to all the rivers of Sicily.

Past writers on the subject have differed as to whether these glittering fossils had actually originated on the island, and whether it was distinct from the true Baltic amber in formation and composition. Ornaments discovered in Italo-Greek and Etruscan tombs nearly always seem to be of a dark red colour, quite outside the normal colour range of Baltic amber. However it could be misleading to determine the origin by colour alone, as the effects of ageing and external influences can darken the colour, affect the quality of amber and produce the patina which is so often seen in art objects of German origin that date from the fifteenth and sixteenth centuries. Modern authorities seem to agree that both Sicilian and Rumanian amber differ in their chemical composition from the Baltic "succinite".

Towards the end of the seventeenth century, amber was held in such high esteem in Sicily that a necklace of it was always mentioned among the presents given to a daughter on the occasion of her marriage. A magnificent present indeed, considering that it was also said to be dearer than diamonds.

Sicily no longer produces amber in marketable quantities, and any green or blue pieces that are found are exceedingly rare and costly. There is no doubt that the colour of some of these pieces is among the loveliest imaginable. I have seen some that resembled flashing opals, and others that were of such a deep blue that they could have been mistaken for lapis lazuli.

The kind of amber which is found in Burma is obtained entirely by mining. This highly fluorescent amber is only partly clear, and the colour varies from red to brown. There is one rather special brown which has the kind of bloom on the surface usually associated with blue Sicilian amber and is sprinkled with tiny black specks resembling tortoiseshell. Burmese amber has always been very popular with the Chinese, who used it for the adornment of their women and to drape in strings ceremonially over the robes of their mandarins.

The ancient Chinese distinguished four different types of amber and gave them different names. The first type was pale brown, clear and transparent and was called *Ching Peh* or golden amber. *Hu peh* or tiger amber was a darker brown; this was believed to receive the souls of tigers after their deaths and was consequently valued highly, as the tiger is second only to the dragon in Chinese mythology. The third type was *Mi La* or honey amber, this was yellow and opaque. The last sort was known as *Chio Naio* or bird's brain and was darker in colour and also opaque.

It is very doubtful whether any Burmese amber is found today. Even in earlier centuries, it was never found in sufficient quantities to merit any real commercial enterprise, but was mined in primitive fashion by the peasants. The fact that so little has been produced makes it very precious indeed and causes me, on reflection, to wish that I had not parted with some of the fine Burmese carvings that I once had in my possession. These were very large pieces which had been sent to China where they had been intricately carved back and front with all kinds of scenes depicting peasant life, showing the peasants working in the fields and in the forests, with the figures of the men and women beautifully marked so that every detail of their clothing was clearly visible. Almost no examples of animal or plant life have been found in Burmese amber.

The third very rare amber is vividly different from any

Yellow-brown and mottled snuff bottles, Chinese, 64mm and 61mm.

Mottled browns and golden-brown snuff bottles, Chinese, 60mm and 64mm.

other. Rumanian amber is the most exciting and unusual variety of all. It was first discovered and mined in a stratum of sandstone close to the River Buzău in Rumania. After great storms had torn away huge chunks of the river banks, thus exposing the geological strata, deep holes were left into which the peasants would dig until they came across blocks of a black coal-like substance which had the lumps of amber embedded in it. This again was mining carried out in the most primitive way. The texture of Rumanian amber is quite soft and brittle, and it is full of tiny fractures that are invisible to the naked eye, so that it does not lend itself to carving.

The colours of this "Rumanite" as it is sometimes called, are quite extraordinary. The dark ones are the most impressive. Gasses absorbed by the amber have caused fractures which reflect light and give a density that intensifies the effect of glowing lights trapped within the amber, making it glitter like spangles. There are also many lighter shades of brown, and a lovely smoky grey; these have a lustre that is fluorescent, and some reflect the light like mother-of-pearl. Although there have been a few pieces of Rumanian amber found which have been almost completely black, it is quite misleading to apply the description "black amber" to the darker kind. This rarest of ambers has shown the most varied and beautiful colours ever to be seen in a fossil resin. Together with the dark background characteristic of Rumanite, all shades of red, ruby red, dark red, brown, green, coffee, chocolate, yellow and black are associated with it.

I believe it is true to say that the quantity of Rumanian amber that has been produced at all times has been less than any other kind, and production today is almost non-existent. Without doubt the three kinds of amber described in this chapter are the most rare, the most precious and the most difficult to obtain.

11. Sicilian green amber necklace, about 90 cm long. Sicilian amber sculpture of a lizard, carved in China, 80 mm long.

12. *Overleaf:* Chinese carved pendant showing a seated lady **writing**, 64 mm high.

CHAPTER TEN

THE TRUE AND
THE FALSE

There are many imitations of amber. The properties, processing and working of these differ according to the ingredients and composition. There is often a loss of the fine polish that the genuine fossil resin shows as a result, but several of the pseudo-ambers actually become more durable in their man-made form. Some exhibit the electrical action which attracts tiny particles of paper and hairs to cling to them after they have been rubbed on a sleeve or some kind of cloth. However, with the exception of "ambroid", none of them exudes the characteristic odour of pine when rubbed on an abrasive surface or set alight.

Ambroid is the nearest thing to real amber that exists among the imitations, because it is made from actual particles of amber which are too small to be used in any other way. These crumbs are the surplus from the working of the larger and more important lumps of amber. Since

1881, the method used to produce ambroid — or "pressed amber" as it is sometimes called — is as follows. The tiny pieces are first softened by being heated at a temperature of roughly 185°F/85°C. This does not cause the amber to disintegrate altogether, but it attains a degree of pliancy and a consistency similar to that of india rubber. In this elastic state it is put into a very strong tray which has a perforated cover. At a higher temperature than before, something in the region of 430°F/200°C, the tray and its cover are pressed together. The softened amber is pushed up through the holes, and then left to cool until it solidifies.

The resultant "pressed amber" can now be turned on a lathe and cut and polished into many useful and decorative articles, though it is not used for any of the fine jewellery or artistic carving in the same way as true amber. Necklaces of ambroid are produced, but these are easily recognisable to an experienced eye as the composition has a certain misty look with what look like rows of tiny feathers underneath the surface. Any other marking in the imitation is seen to be much more regular than in true amber, which is often full of all kinds of irregularities and imperfections.

Although ambroid is stronger than amber and has a rather elegant look about it, it does not improve with wear, whereas the original resin becomes even more beautiful during use, especially in the case of bead necklaces. The composition has been fashioned into such objects as walking-stick handles, door knobs, fancy handles for ladies' umbrellas; all fairly bulky articles which would require a very large piece of amber to make. The finished thing would be a very costly item indeed, even supposing it were possible to obtain the amber to do it. However it is in the smokers' world that pressed amber has really come into its own. One sees all shapes and sizes of cigarette and cigar holders, also pipes with a mouthpiece of this material. These have often been teamed up with a bowl made of meerschaum or ivory.

In the early years of this century, quantities of bead necklaces were produced which resembled amber but were actually made from that compound of phenol or cresol plus formaldehyde which is known as "Bakelite". Of course this contains no amber whatsoever and has no odour of pine, but it does possess some electrical power similar to amber. The beads were nearly always red or brown in colour and very highly polished. A great many of these necklaces are still in existence, treasured family heirlooms faithfully handed down in the firm belief that they were amber.

With a material so much prized through the ages, it is not surprising that some of the techniques of faking go back a very long way. The Romans of Pliny's time despised the white amber and used it only for its aromatic qualities, as a sort of incense. However industrious Roman craftsmen were

A newt preserved in Kauri gum or Copal, New Zealand, 96mm across.

already able, by the judicious use of plant or animal dyes, to produce a range of more valuable colourings from the pale stone. More recently, in Conan Doyle's story *A Study in Scarlet*, Sherlock Holmes mentions a whole industry which existed for the purpose of putting fake flies into fake amber to make it look truly convincing.

Among the substances which have been used to imitate amber — and these include even ones with widely divergent qualities, such as glass and celluloid — there are found numerous different gums, resins and mixtures of these with other ingredients, both natural and man-made. This can be very confusing to any but the most experienced. To a practised eye, these other resinous substances could not possibly be mistaken for amber, though many of them are actually of fossil origin. However the trees that produced them are, in most cases, unidentified. Occasionally small inclusions have indicated that the parent tree was some kind of pine, but not the amber pine. I can only attempt to name a few in this chapter.

Glessite is a resin from an unidentified tree, light in colour and of a rather sticky texture. Gedanite is known to come from a species of pine. It is very brittle and difficult to work on, but has sometimes been found to contain small insects. Its colour is a murky yellow, partly clear and partly opaque.

An interesting resin which originates in New Zealand, is known as "Kauri gum". This gum is the secretion of the Kauri pine which still grows in New Zealand, and it is therefore a much more recent product than amber. Some of its characteristics are very close to amber indeed. It has, for instance, the same density (1.08) and the same refractive index (1.54). When burnt, it has the same sweet aromatic smell, but it burns much more readily than amber. Kauri, or Copal as it is sometimes called, may be brown or yellow in hue, but is more usually almost colourless, resembling dirty water. It incorporates many tiny bubbles and is impregnated with odd dark brown marks all over the outside surface. Unlike amber it can be melted down and poured over plant and animal specimens, to preserve them, when set, as an illustration and object of scientific observation for students of natural history. I have in my possession a fine specimen of a newt thus displayed in Kauri gum.

Two other similar gums about which very little is known are Beckerite and Stantinate. The first, a dull brown in colour, is noted for its unusual toughness: it is almost impossible to work with. The second, Stantinate, is a more attractive colour, a brighter reddish brown, and softer in texture. Both these resins were found in the same ancient forests that produced *Pinus succinifera*.

There are many more, but it is not possible to analyse all of them. Some of them are natural substances, others are man-made. My uncle, who spent a lifetime working on amber, could tell at a glance the composition of many of the pseudo-ambers which were often brought in to us to be identified. It amused me to discover that one of them actually included casein, the protein found in cheese, among its components.

All these compositions are extremely difficult to process, and in my opinion hardly worthwhile, as none of them can really compare with the original resin which has survived so many natural processes during its incredibly long lifetime. As a result of this slow maturing, genuine amber has a beauty and a mystery of its own which can never be faithfully reproduced.

This chapter on the pseudo-ambers would not be

complete without some reference to a material which has often been confused with amber, but is actually in no way connected with it. This is the substance known as Ambergris, a secretion found in the intestines of whales which has been used as a base for perfumes for many years. Before being cleaned and treated, ambergris gives off an unpleasant odour, but this decreases as it dries and hardens. Strangely, when a small amount of it is mixed with other essences, it has the effect of strengthening their aromas while its own goes unnoticed. Ambergris is a soft light material and can be taken internally. Indeed, in the late sixteenth century it was regarded as a potent drug and much valued as an aphrodisiac.

One other process worth mentioning here is the manufacture of various kinds of amber varnish. The basic material for this is the mass of white powder caught in large trays placed underneath any machine or lathe engaged in the cutting of amber. This is collected and sold to manufacturers to be made into varnish. This varnish is very valuable for covering all kinds of wood surfaces; one of its most important uses has been for varnishing fine violins. The making of amber varnish should never be attempted by persons who are not used to handling it, as it can be extremely dangerous. The softened amber powder is mixed with boiling oil over a fire, then carefully removed and left to cool. When quite cool, it is thinned out with measured quantities of turpentine, and applied as necessary.

We come now to some ways and means of testing amber. Many of these are inconclusive and others can result in damage to a treasured possession so that great care must be taken in carrying them out.

The basic tests for amber are its density — it will float in salt water — its aromatic smell, and its frangibility. To test

A shrine in amber and ivory, German, 17th century.

the resin by burning, the object in question must come in contact with a small but very hot tool, such as a needlepoint, then immediately be held as near the nostrils as possible. The burning resin gives off deeply aromatic odours that are both mysteriously soothing and at the same time very exciting. It is an extraordinary thought that this pungent smell comes to you as strongly as ever after the passing of millions of years, the same scent that emanated from the original pine tree that was the source of its existence. Amber will not disintegrate entirely and only melts at an extremely high temperature.

If the burning results in other strange odours, the substance being tested might contain casein — the protein obtained from milk — which would give off a smell of burnt milk or cheese, or it might be Bakelite, which smells strongly of phenol (carbolic acid).

A natural lump of amber, set as a pendant in gold, 42mm × 25mm.

If amber is rubbed with a little ether, the latter will have no effect on it, but on copal resin it acts as a solvent and the rubbed surface will become tacky.

As regards frangibility, an inconspicuous part of an ornament can be selected (e.g. the drill hole of a bead) and the surface can be lightly scraped with a penknife. If the substance is amber, it will break away sharply with a shuddering motion. If it is plastic it will peel off, being sectile*. Bakelite is very tough and usually resists this method of testing.

The use of ultra-violet light (long wave) can also be useful in judging amber. In general it fluoresces a mustardy yellow colour, but if a small scrape is made, the freshly scraped surface will exhibit a blue fluorescent glow.

Amber develops a charge of static electricity after being rubbed on a sleeve or other material handy, and it will then attract or pick up and hold small pieces of tissue paper or other tiny fibres. This is not a conclusive test, as most imitation ambers will do the same. However, if it does not pick up at all there would be strong reason to doubt its authenticity.

Bearing all these tests in mind, it is as well to remember that one of the most vital characteristics to look for when buying amber is the weight. There are no two ways about the fact that it is always very light compared with other minerals. Therefore, if beads or other articles are heavy, it can be safely concluded that they are not genuine.

There are comparatively few people in this country familiar with the working of amber, but those who are will know that it is essential to choose the right type and quality of piece according to the use it will be put to. For instance,

*Capable of being cut with a knife, without breaking.

if a block of amber is required for a mouthpiece on a pipe it must be as flawless as possible in order to avoid a breakage. Oddly enough, amber which has been taken from the sea is usually stronger than that which has been mined, though the latter is often of a finer colour.

A mouthpiece to go on a pipe frequently has to be curved. The piece must be soaked in oil for a while to prevent the surface of the amber from drying up when heat is applied, and then gradually warmed over a gentle flame and bent into shape with the greatest of care. Whenever bending has to be done, it is most successful if the amber is of the absolutely first quality. All rough amber is first cut into the required shape with a special saw, and then reduced to its final form by filing, scraping or turning on a lathe. It is finished by polishing with a little spirit of wine and Vienna chalk on a machine, using first a rough buffer and then a soft one which gives the final lustre.

In conclusion, and speaking from long and close acquaintance with this gem, I would venture to say that it can never be an easy subject to work on. The men who do carve, shape and polish it into luxurious, beautiful things that give pleasure to so many, can be classed among the most artistic craftsmen of our time.

CHAPTER ELEVEN

RECOLLECTIONS

At one period before the Second World War, Sac Frères was situated in the Burlington Arcade, near enough to Burlington Street itself. It was in fine company among that elegant collection of small shops which offered for sale all kinds of precious gems, fine linen, china, leather and cashmere: an endless selection of luxury goods for those whose pocket books could afford to buy them.

The Amber Shop still remains in the same area where it has been for so many years, tucked away at the very end of London's Old Bond Street, and I should like to devote this chapter to the essential character of the shop and to its founder, my father, who died in 1966. His name, Arthur Charatan, is still on the door in somewhat faded letters, and he is still remembered by the many people who came into the shop not only to buy, but who liked to sit down and chat to him.

In an article in *The Observer* on 12 June 1966, it was said that people enjoyed buying amber from him because they were "soothed and entertained by his mournfully humorous conversation". This was very true, equally characteristic was his ability to tell a good story spiced with his particular brand of wry humour. He was greatly respected for his highly specialised knowledge of amber, which for him was the main interest of a long lifetime. It is a fact that for many tourists visiting London, "The Amber Shop" was the first stop on their itinerary.

The shop in my father's time looked as though it was a film set representing a scene from the works of Charles Dickens. The lighting was dim and, though the interior of the shop was very tastefully fitted out with showcases and cupboards of beautiful dark mahogany, the condition of it was such that it would not have been too difficult for a client upon entering to imagine that he or she had indeed wandered into *The Old Curiosity Shop.*

Once inside, they would invariably find my father sitting beside an old and extremely lovely table of Italian origin. This table, which is still in our possession though no longer in the shop, is oblong in shape and supported by four satyrs, their bearded heads pointing outwards and their cloven hoofs forming the legs. This same table, however, was the cause of constant friction between my father and myself, and indeed any other tidy-minded person who happened to be working in the shop. He kept it piled high with dusty papers and books which accumulated week by week. No one was allowed to disturb this sacred heap except himself, and oddly enough he could always put his hand on anything that was needed with no trouble at all.

There are many people, too, who will remember my uncle Sydney Charatan, my father's brother, who worked in close

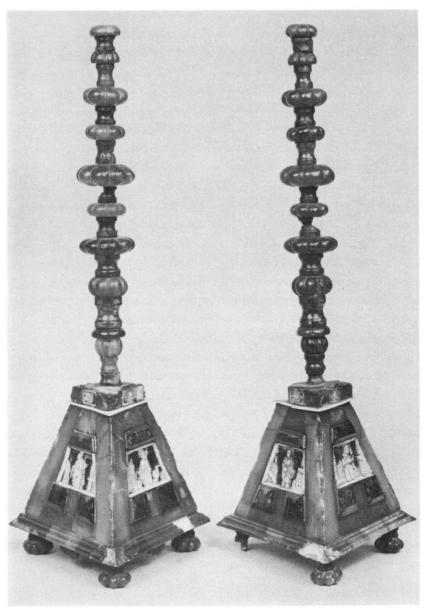

Pair of amber and bone candlesticks, German, 315mm high, c. 1700.

association with him for over fifty years, and who survived him by just three years. His work kept him more or less permanently out of sight, downstairs beneath the shop in what we called "the dungeon". Here he carried out the practical craft side of the business, cutting, polishing and repairing articles of amber. He could repair the most fragile pieces, handling them with the utmost delicacy. When he did come up the stairs into the shop to discuss some problem with his brother, or to talk to a particular customer, he could not avoid looking a little like a snowman as the white dust, which flew everywhere from the amber he was working on, covered him from head to foot. After he died, there was no one left who could equal his ability and specialised knowledge in the working of amber. It was no longer possible to give people the service that we had been able to in repairing broken pieces, as we had done in my uncle's lifetime.

Looking back over a long span of years of specialising in this one subject, it was inevitable that there were good times and some very bad times. At one period trade was very bad indeed and there was little work for any of us to do. We were at that time in a small shop just off Bond Street, not in the street itself. Day after day we sat downstairs in a little room which was used as a kitchen, hoping that things would improve; but I could see that my father was very worried. I was then in my early twenties, full of ideas for promoting trade to help us out of our difficulties, and found myself impatient and surprised that he chose a "wait and see" policy and seemed content to do just that. He was right, however; after an anxious period, business did improve, and soon things were back to normal and we were as busy as ever.

An amusing visual record of that six months' waiting was

Amber and ivory casket, German, 18th century.

that a whole wall of the little kitchen room under that particular shop became filled from end to end with rather splendid sketches. To while away the time my father had covered the wall with drawings of bearded Chinamen, horses, elephants, Buddhas, dogs, cats and rabbits, and even what I thought a very good attempt at a self-portrait. He was a very philosophical man.

Working in The Amber Shop, especially from 1966 onwards, I have met many customers old and new who have a great feeling for amber. We are extremely fortunate in having a clientele of interesting and charming people who still find pleasure in visiting us and sitting down to chat for half an hour, as they did in the old days when my father was sitting by his chaotic paper castle.

We count many Americans among our amber lovers, and it was very gratifying and amusing to hear that one small section of the inhabitants of Boston has taken to having

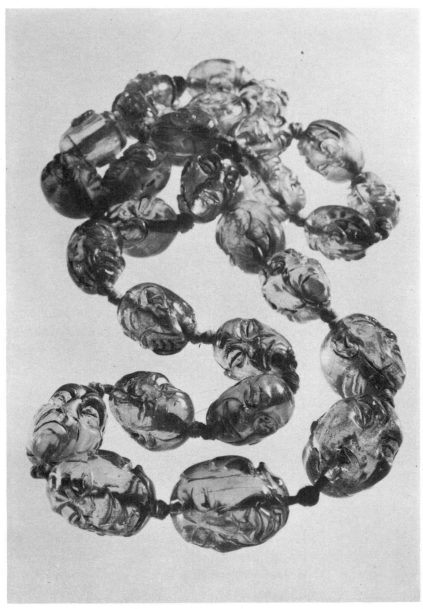

30 bead, red amber necklace, each bead carved as a head; Chinese, 700mm long.

13. Baltic amber necklace with oval beads, about 110 cm long. Two modern rings set in silver, the larger is 62 mm long.

14. *Overleaf:* Chinese carved snuff bottle, 85 mm high, with amethyst stopper. The side shown depicts a mandarin sitting under a tree, the other side shows a branch of cherry blossom.

"amber parties". When one member of a circle has been to Europe and brought back a beautiful piece of amber jewellery, all the friends and neighbours, wearing their own treasured amber, gather to celebrate the new acquisition.

It is not possible for every person to do the work he or she enjoys, so I count myself very fortunate in this respect. I fully believe in the mystic properties of amber, it is such a good feeling to be surrounded by great masses of it, to handle it, and to be able to inhale its subtle odour. If you possess any amber at all, it is very likely to be the oldest thing you have, but in addition it is a living thing. If it is well cared for, it matures and mellows and with time becomes even more beautiful.

About fifty years ago we purchased an enormous quantity of a fine silk from China. We have always used this special cord to thread our amber necklaces, usually putting small knots in between the beads, which prevents them from rubbing against each other and ensures that, in case of accident, only one bead will fall when the thread is broken. Unfortunately today we have only a relatively small quantity of the silk left, enough to go along with, but the colours are now somewhat limited. It is fascinating, however, to look upon the now dirty and musty packets with their faded Chinese writing, and to realise what a different world exists today for that country and its people.

Anyone who has visited us over the years has always found one or other of the family in attendance. In fact we have been given the charming pseudonym of "The Amber Family" by some of our customers. It is sad that such small specialist shops are few and far between in today's world of large companies, and I hope that "The Amber Shop" will continue to flourish among the giants for some time to come.

During the past decades, there have been many superb examples of this jewel from the sea that have passed through my hands. Some of these were so outstanding that it would not be difficult to picture them for you here, and this is what I should like to try to do.

In a business which handles only articles in amber, we have seen many unusual antique rings, and I would like to describe a very special one which was reputed to have belonged to Lucrezia Borgia. It was a round piece of fiery red amber, about 1 inch/2.5cm in diameter, and most intricately carved into the curved body of a snake, the head raised with the forked tongue poised to strike. This was mounted on a plain gold band with six claws holding the body of the snake firmly in place. Quite apart from the beauty of the amber and the intricacy of the carving, this was also a poison ring. The whole surface of the ring could be raised by touching a minute round knob located on the underside. Presumably a tiny phial of some poison could be inserted into the space inside the ring. Whether this was meant for the wearer or some other luckless victim is a matter for speculation: either way the ring itself was all a collector of unique and lovely things could have desired.

It was while the shop was in Burlington Arcade that my father and I made the acquaintance of a collector of antiques of many kinds. His name was Thomas Kitson, of Whitehall Court in London's West End. His home was a treasure-house of beautiful and valuable things, and his amber collection was one of the finest in the world. In particular, he owned a superb collection of necklaces which he had obtained from a variety of sources, one of which was our shop. He kept them in a tall antique chest containing a great many drawers running from top to bottom on both sides.

Whenever we dined with this genial gentleman, now unfortunately deceased, he gave us a mutual treat as it was his habit, after the meal was finished, to go right through the contents of the chest, so that we could examine and enjoy the sight of these beautiful beads together. It is not possible to remember every one, but certain gems in the collection were unique.

The chest contained several fine and typical mandarin necklaces. In these the amber beads are invariably round in shape and clear, the colours light or dark gold, brown or red. The necklaces are very long and there is no graduation. They are ornamented with beautiful pieces of jade or ivory which hang from the sides and the middle.

There were very many necklaces of beads that had been carved in China. One of these was graduated, with a dozen very large beads towards the middle. Each bead showed the head of a mandarin and every face had been given a different expression. Some were smiling, some grimacing, others were a solemn or a sad look. There were grim faces, wise faces, faces adorned with long beards. Many wore curious hats, and all were exquisitely carved.

Another unusual and fascinating necklace in which the amber was golden in colour, but was partly clear and partly opaque, had been carved with the different signs of the zodiac on every bead.

Among the most rare and colourful necklaces were those that came originally from Sicily or Rumania. The amber that is called "Blue Sicilian" is rare indeed. The colour is not a true blue but is soft and plumlike with overtones that appear and disappear like mischievous and elusive sprites.

Rumanian amber has a very concentrated crystallisation and the colours are mainly either a light brown or an even rarer black. It is flecked with gold in either case. In the

Kitson collection there was one black Rumanian necklace of such intense colouring that it glittered as if encasing a living form.

Apart from this and other collections, there were certain other pieces that came our way which were unique. One of the most striking was a carved Chinese Buddha. The colour of this piece was quite extraordinary. Looking at it on one level it was completely black, and on another it was a perfect green. As it sat upon a clear golden stool of amber the effect was one of incredible beauty.

At one time we possessed a magnificent set of carvings representing the eight Chinese Immortals. These were of dark brown opaque amber and stood 5in/12.7cm high without their stands of carved black wood. They, too, were a most impressive sight.

It is very rare today to find an antique chess set in amber. We were very fortunate indeed as we were able to buy a sixteenth century set which combined both chess and the game of draughts. It had belonged to an elderly Baroness of German origin and the set had been in her family for hundreds of years, handed down to the heirs century after century, until this lady reluctantly decided to part with it. The draughts side of the board was more in ivory than in amber, but the playing course was defined by inlaid strips of amber richly decorated by a technique known as Verre Eglomisée or painting on the underside of glass. Parts of the miniature reproduction are painted in colour, and other parts are engraved as in an etching on a ground of gilt or silver, so that when finally they are picked out in black lacquer they appear as black hatching on a gold or silver background. The whole is finished off by a backing of silver foil which sets off the brilliance of this decorative technique. This specialised work was widely used on amber, and was a

German toilet casket/cabinet with 5 drawers and 10 bottles, 250mm wide, 17th century.

substitute for enamel painting. It appears to have been taken over from glass technology and was much in use in the sixteenth century for representing coats-of-arms on the base of glass beakers. During the sixteenth and seventeenth centuries, the amber workers frequently signed their work in this medium. In this particular set the chess side of the board had been worked in alternate squares of clear and opaque amber, with inlaid reliefs alternating with Latin inscriptions. This beautiful set was soon purchased by a collector, and I have never seen anything to match it since.

We handled many antique jewel caskets, mostly finely worked in the incrustation technique so brilliantly illustrated in the fabulous "Amber Room" at the Tsarskoe Selo Palace of the Tsars of Russia. Plaques of amber, delicately carved, alternated with similar ones of pale ivory. We had one which was of special historical interest as it

Dark golden amber snuff bottle carved with Buddhist motifs, Chinese.

came from Hever Castle and had belonged to the ill-fated Anne Boleyn, second wife of Henry VIII. The casket was very old and fragile, and had many small drawers lined with dark green velvet which exuded a fascinating odour of musk. Given voice, this casket could have told a dramatic story of the precious jewels it had contained when its owner had been high in the favour of that great king of England and the tragedy which followed when its owner suffered a violent death on the orders of that same royal husband.

Few collectors of snuff bottles could have failed to appreciate the fine examples that came our way from time to time. They were made in all kinds of amber and all different colours, some perfectly smooth and opaque, others exquisitely carved by patient Chinese hands. Looking at one of these bottles decorated all over the surface with figures of mandarins with birds, trees and flowers, it is difficult to estimate the time, the care and the patience it must have taken to achieve this work of art. One slip of the tool in the artist's hand could have ruined the work of years. Practically all snuff bottles have small stoppers with an ivory spoon attached to them; they make a pleasing contrast to the amber in various other gemstones such as jade, amethyst, tourmaline, ivory, turquoise and many others.

I would like to mention here that, in the opinion of many writers of natural history, it is not difficult to cut and carve amber. From my personal experience with members of my family who were themselves artists in this work, I cannot agree. It is of the utmost importance in the process of working amber that the operator has considerable knowledge of the subject. Only the most practised eye can distinguish with any certainty the quality of amber which lends itself to treatment from that which cannot be dealt with safely. The slightest speck or defect, almost invisible to the unaccustomed eye, may be sufficient to cause the piece to break at the first attempt to cut it. This can be very costly in terms of both labour and material, and causes a great deal of frustration to the artist.

Having endeavoured to picture some of the rare articles in amber that were carved and shaped by hand, I cannot bring this chapter to a close without some reference to the numerous pieces that need no human hand to embellish them. These are the natural products of the earth and the sea which have been picked up from various places all over the world and left in exactly the shape and form in which they were found.

As a collector of amber, I find these untouched pieces among the most intriguing, and am fortunate in owning a few that are outstanding of their kind. I have a small piece of clear amber which was found on the coast of Sicily. Its beauty and charm lies in its amazing texture and colour. The first attribute is apparent as it lies in the hand. It gives the impression of being not a fossil, but a smooth seductive piece of silk. The colour is dazzling as one beholds the clear lustrous gold centred by a flash of brilliant red, almost as if during its formation the resin had captured a sunset as the end of a hot summer's day.

A second and much larger piece, which was picked up in England on the Suffolk coast holds the interest in a different way. The contents of this pale clear yellow lump seem to constitute a small underwater world. In its liquid state this single piece of amber caught and held many extinct species of plant life, fragments of wood, two small mosquitoes, pine needles and several small flies. All these have been preserved for millions of years to be studied and enjoyed by nature lovers all over the world.

Among the rough uncut pieces that merit a special place

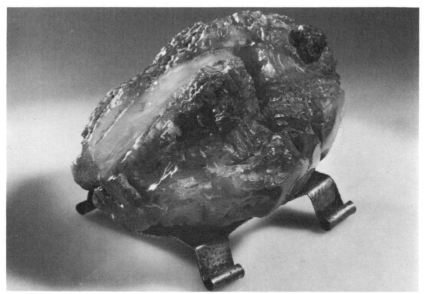

Large opaque, unpolished lump of Persian amber, 180 × 115 × 105mm.

in my small collection are several that came originally from Burma. These were not obtained from any sea shore, but were mined from the reddish-coloured soil wherever small veins had been discovered and investigated. The soil around these diggings when freshly dug emitted a very pleasant scent. This Burmese amber reveals a great depth of colour varying from a dark brown to red, with a texture that is soft and smooth to the touch.

Of great interest to collectors are the various lumps of opaque nut-brown Persian amber, some still covered with the rock-like crust which is their outer covering before they are stripped and polished.

The numerous other unusual colours that are found can be the consequence of climatic conditions in certain countries or due to the soil in which the amber was buried for countless centuries. Some pieces that have been immersed in water are covered by tiny crustaceans.

111

CHAPTER TWELVE

THE DISPLAY AND CARE OF AMBER

B efore concluding, I should like to give the owners of any kind of amber a few simple hints on how to take care of it. If you own any carvings, snuff bottles, netsukes or other similar objets d'art, being very precious and possibly fragile, these need to be handled with special care to prevent chipping or breaking. It is best to keep them inside a glass-fronted showcase so that they can be viewed in safety. They should be gently dusted with a soft cloth from time to time.

The same treatment should be given to amber jewellery. This too only needs a small amount of gentle polishing with a clear duster if it becomes dulled through being handled or exposed to a steamy atmosphere. Be warned, however, that the hands should never be immersed in water when wearing rings of amber.

The things which, in my experience, are most often exposed to careless usage are earrings and necklaces. This is mainly because they are allowed to come into contact with liquids such as perfume and hair sprays. If beads or earrings become covered with fluids of this type, the gloss can be permanently dulled and they are spoiled for ever.

Amber beads can be washed in warm soapy water if they have become really dirty, but this should only be done if it is absolutely necessary. After they have been dried, a small amount of clear olive oil may be applied and then rubbed off to restore the polish.

Finally, I would like to say to those who, like myself, are lovers of amber: wear your necklaces and jewellery as much as possible. Amber does not take kindly to being shut away from sight in a covered box, its texture is more likely to improve and darken when adorning the necks of wives and daughters, and it can acquire a beautiful lustre from being handled and from close contact with the skin.

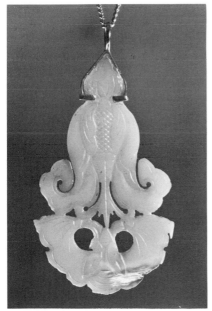

Amber pendant carved as stylised flower, Chinese, 49mm high.

113

CHAPTER THIRTEEN

THE LITERATURE AND MYTHOLOGY OF AMBER

There are many references to amber in English literature, both in poetry and in prose. Not many authors, however, have chosen to make it the pivot of an entire work of fiction, as did Hugh Walpole in his novel *The Old Ladies.* Three ladies, united only by their circumstances of extreme poverty, decrepit old age and loneliness, live on the top floor of a grim old house in a decaying square of a cathedral town. The poorest, loneliest and most timorous of the three — the newcomer; plain, awkward, absurd, pathetic Miss May Beringer — owns a magnificent piece of carved red amber which was given to her by the one great friend and love of her life, Jane Betts, long since married and gone away.

"'It will warm you, my dear,' she said when she gave it to her, 'always keep you warm like my affection. Never lose it or sell it. My heart is inside it!'

"So did this chunk of amber enshrine both their affections. It was shaped square like a little block of wood, and this block was surmounted with a carved red amber dragon. It had in it the most lovely lights and colours, that flashed and trembled from the deepest Venetian red to the fairest honey gold..."

To the second old lady, gentle proud Mrs Amorest, the amber is simply the most beautiful thing she has ever seen. "When she held it to the light and saw the shaft of gold strike through to its very heart, when she saw the liquid bubbles of rich ruby red that danced in the cleft of thick, honey-coloured, misted fibre, when she saw the dragon with his flaming head and gold-flashing claws, when she felt its sturdiness and independence and form, she could only say and exclaim, as she replaced it reverently on the mantelpiece: 'You *are* fortunate to have it! It lights up all the room.'"

But to the third old lady, heavy, sullen Agatha Payne with her gipsy looks, her sweets, her cards and her passion for bright colours, the amber becomes a terrifying obsessional object of desire.

"She saw . . . as though it had been placed there for her especial glory, the heart and centre of all the colour in the world... She felt, as she held it, that it was hers, that it had always been hers for ever and for ever and for ever."

In her consuming lust and determination to possess this wonderful thing, she coaxes, bullies, torments and terrifies poor Miss Beringer literally out of her life. It is an eerie, powerful story that haunts the memory, and Walpole has done full justice to the magic of amber.

Its colour and glow have often been invoked by poets, the very word used to convey the idea of a clear liquid gold. Among the writers who have used it thus to enhance their

One side of a Chinese snuff bottle showing a rat among rocks.

imagery were Shakespeare, Burns, Pope, Tennyson, Carlyle and many others. Perhaps the most flattering reference of all occurs in Milton, who showed it great honour when he adorned the "Chariot of the Messiah" with the golden stone (*Paradise Lost, Book V*).

Another work of fiction relating to our subject is *The Amber Witch* by Wilhelm Meinhold, which was translated into English by Lady Duff Gordon. It tells the story of the trial of a young girl for witchcraft.

The scene is laid in a small German town at the mouth of the Oder during the Thirty Years' War, in the first half of the seventeenth century. We meet the chief characters just after the Imperial troops had departed from the district with all the food and valuables they could lay their hands on, leaving the country destitute for miles around. Pastor Schwiedler, a rigorous old Protestant, tells of the great need and hunger of his whole parish, including his own

The other side of the same bottle showing a rabbit eating a melon, 69mm × 44mm.

household which consisted of himself, a daughter in her teens called Mary, and an old maidservant.

One day Mary went to pick blackberries near the sea-shore. To her amazement she found a rich vein of amber hidden from sight under a pile of dried leaves and twigs. Filling her apron full she hurried home to show it to her father, after carefully concealing the remainder again. The pieces were very large, two of them nearly the size of a man's head. The pastor was very excited and, later that night, he accompanied his daughter to where the rest was hidden. They filled a sack with all the amber they could find and then returned home. They were able to sell the entire find to a Dutch merchant for 500 florins. This wealth allowed this admirable family to keep every house in the parish supplied with food until the famine was over.

That winter a second piece of good luck followed the first, as another large quantity of amber was discovered. This time it was washed up on the beach during a violent storm. This enabled the villagers to restock their fields with cattle and sheep, and all seemed well once more.

Just as the community was recovering from the destruction caused by the soldiers, some of the cattle were stricken with a mysterious illness; according to the custom of the time, this was attributed to witchcraft. Mary, following a superstition of the period, was often called upon to attempt a folk cure. Unfortunately the cure only appeared to succeed in a very few cases, and many died.

As is the case with many genuinely good people, Mary had her enemies, and they soon took advantage of her failure to cure the animals, and accused her of witchcraft.

Finally, at the instigation of the villainous sheriff, Mary was arrested and charged with witchcraft. After repeatedly denying her guilt, she was put to the torture. Rather than endure so much, when she knew she could not hold out under such pain, she confessed to everything. The poor girl was condemned to be burnt at the stake, the execution to be carried out at dawn on the following morning.

The next day the terrified girl began the long walk to the place of execution. She was accompanied by the sheriff and his officers. When they came to a bridge over a stream, the sheriff on his horse was the first to attempt to cross it, but as the animal set foot on the bridge it was as though the planks were covered in thick and slippery ice: neither the sheriff and his horse nor anyone else who endeavoured to gain a foothold could do so. Bewildered and frightened at this strange development, they conferred among themselves: this could only be the work of witches, and particularly the one who walked in their midst. In fact, the only bewitching the bridge had undergone had been a coat of tallow well pasted on by the miller's son.

15. Three natural pieces of amber: the clear gold piece has an opaque cloud inclusion, 99 mm long; the piece at top centre is set in a seal, 45 mm high, and the third piece is Sicilian and is carved as a Mongolian head.

16. Chinese 'Mandarin' carved head necklace in opaque amber, about 100 cm long, with matching silk tassel.

Amber and ivory crucifix, German, c. 1700.

Impatient at the delay, the sheriff fiercely spurred his horse once more on to the bridge. This time he succeeded in reaching the middle. As he did so, there was a great clap of thunder and a flash of lightning which caused the horse to rear sharply. The sheriff, losing his balance, was flung over the parapet on to a great mill-wheel below, and killed outright.

The others decided at last to avoid the bridge and take another way across the fields, and so the march to the stake continued.

When they were almost in sight of the dreaded place, a single horseman was seen galloping up behind them. It was Mary's sweetheart, Rudiger von Nieukerken.

The constable pulled out his dagger to stab the helpless girl, but the young lord had already reached them and was too quick for him. The constable fell, pierced through the chest. He confessed that he had known all the time of the plot to have Mary condemned as a witch, but had kept quiet on account of a large bribe Baron Wittich had given him, though he knew the girl was innocent.

Saved from the horrible fate, Mary fell into the arms of her lover, while the people of the parish drifted away in shame and belated condemnation of their own cruel behaviour.

Although there is no particular moral to be gained from this tale involving amber and witchcraft, it emphasises the fact that amber was not only an article of luxury in the seventeenth century, but of such economic importance that it could be the means of solving a financial crisis for an entire community of the time.

The Greek myths, like the myths of all the nations of the world, represent the ancient beliefs of primitive man, his attempts to explain the phenomena of nature. The problem of the origin of amber was always a matter of great interest, from the earliest ages. Perhaps there is no comparable substance which has produced so many conflicting theories.

One well-known Greek legend concerning the origin of amber is of special interest, and has all the quality of intense poetic imagination characteristic of Greek mythology.

The story tells how Phaeton, the son of Phoebus Apollo the Sun God, entreated his father to trust him for one day to drive the Sun Chariot pulled by the wild horses of the Sun. Phoebus at last consented, much against his will, to his son's request. The horses were harnessed to the chariot, and Phaeton started on his way. For a time everything went well, then suddenly and for no apparent reason the horses took fright and bolted. Phaeton was unable to control them, and coming too close to Gaea, the Earth, set her on fire. This was said to have been the origin of volcanoes. Now everything was blazing, the forests were destroyed and the lands everywhere parched and dry; so terrible was the heat that the people of Africa were burnt from white to black. Then Gaea begged the mighty God Zeus to save her, and he responded by killing Phaeton with a thunderbolt. Phaeton's body fell into the River Eridanus (the ancient name for the River Po) and lay there until the nymphs of the stream lifted out the dead charioteer and buried him by the river banks.

After some time his three sisters, the Heliades, came to look for the grave. When they eventually found it, they sat on the ground and wept copious tears. Unable to move away, they stayed there until their wasting bodies took root in the ground and became covered with the bark of the surrounding trees, their arms turned into boughs, and at last they themselves were completely changed into trees. Their tears, however, continued to flow and, as they

Variegated amber casket with applied ornaments, German, c. 1700.

hardened in the heat of the sun, turned into amber. A strange story, and one which was regarded by Pliny with some scepticism in his early writings on natural history.

It is interesting to note how much amber was admired from the earliest days of literature. One famous classical writer, Martial, wrote many charming verses referring to amber in his *Epigrams*. The one I quote below gives a special dignity to the death of a bee enshrined in amber:

Here shines a bee closed in amber tomb
As if interred in her own honey comb.
A fit reward fate to her labours gave,
No other death would she have wished to have.

Another verse by the same author shows the same process of insect preservation and the almost timeless value of amber as a near perfect embalming fluid:

On weeping poplar boughs a viper crawls.
An amber drop upon the reptile falls.
Amazed, she feels the gummy chains around,
But in their hardening mass she's safely bound.
Her royal tomb Cleopatra need not prize,
For in a nobler one a viper lies.

These verses from Martial refer to the poplar tree as the source of amber, following the Phaeton legend. Some ancient writers, such as Pliny, would have agreed that the words "pine tree" would have been a more accurate description of the source of resin, without in any way disrupting the poetic flow.

The Heliades legend related at the beginning of this chapter was perhaps the most famous of all the myths concerning amber, and an elaborate version of it can be found in the works of Ovid.

With this glimpse of the mythical aspects of the "extract from the rays of the sun", I must conclude my story. It has

Necklace of huge beads, Russian, largest bead 40mm across.

followed the magic resin from the lush green forests of the Tertiary Period, through the age of land and trees submerged by glaciers, down to the bottom of the sea, to be washed out again by the action of waves and storms, flung on to beaches, picked up by the eager hands of tribesmen, and used to improve their meagre living standards. We have seen it become the most important gem known to the ancient civilized world, promoting enterprise in trade and establishing new trade routes between distant countries, and stimulating artists and craftsmen through the ages.

Finally we have seen how it inspired the imagination of poets and scientists with such mystic and fanciful conceptions of its origins as the Greek legends reveal, and how its beauty has enriched the work of many of our own English writers.

I end now with my own small tribute to its golden image.

The Story of Amber

I hold a piece of amber in my hand,
Warm to my touch and light.
I lift it up to gaze and stand
Amazed. The sun is bright
But equals not this liquid glow
This pool of light that seems to flow
Through space.

So many million years ago
Where ancient forests grew
The scented resin filtered slow
Downward, gold in hue;
The strange and varied insects caught
In unavailing struggles fought,
Entombed forever.

Then over all the conquering seas
Now covered all the land
And swept away the tall pine trees
With Nature's ruthless hand.
So many treasures buried deep
The mighty waves were still to keep
For centuries.

And thus when land and sea revealed
Their secret, held so long,
This precious thing the depths unsealed
Was amber, light and strong.
The colours glowing red and gold,
A transformation to behold —
A jewel.

Rosa Hunger

125

BIBLIOGRAPHY

Buffum, W.A., *The Tears of the Heliades;* Sampson Low & Co, London, 1898.

Maccall, W. and Haddow, J.G., *Amber: All About It*; Cope & Co, Liverpool, 1891.

Pelka, Dr Otto, *Bernstein*; Richard Carl Schmidt, Berlin, 1920.

Rohde, Dr Alfred, *Bernstein*; Deutscher Vereinfür Kunstwissenschaft, Berlin, 1937.

Williamson, Dr George C., *The Book of Amber;* Ernest Benn Ltd, London, 1932.

Amber, Encyclopedia Brittanica.

INDEX

127